THE FERNS OF THE PLAINS OF TAMIL NADU

THE FERNS OF THE PLAINS OF TAMIL NADU

D. SUBRAMANIAN

Professor of Botany (Retd.),
Annamalai University,
Annamalai Nagar – 608 002,
Tamil Nadu.

MJP PUBLISHERS

Chennai   Trichy   Tirunelveli   NewDelhi

ISBN 978-93-88694-74-2   **MJP Publishers**

All rights reserved   No. 44, Nallathambi Street,
Printed and bound in India   Triplicane, Chennai 600 005

MJP 764   © Publishers, 2020

Publisher: C. Janarthanan

Project Editor: C. Ambica

Preface

The plains of Tamil Nadu from sea-shores of Bay of Bengal up to 500 feet level of various places do not have much favourable climatic conditions to the existence of ferns as in the higher elevations of hill stations like Valparai, Ooty and Kodaikanal. But now-a-days, as the ferns are considered as more important ornamental and medicinal plants, it is important to study and conserve them all over the country. Therefore, the fern flora of the plains has been described in the present book. It will be more useful to the future Scientists and Horticulturists for further investigations in these lines. For the first time, the ferns of the plains of Tamil Nadu have been studied, as far as the author is aware of.

Acknowledgements

The author expresses his sincere thanks to Dr.N.Vijayakumar, Professor of Botany, S.T. Hindu College, Nagercoil for having helped him in providing all the facilities for his stay at Nagercoil for 3 days and taking him in his car along with a helper to various placesalong the way from Nagercoil to Thiruvananthapuram for the collection of ferns, mosses and some of the flowering plants.

He expresses his heartful thanks to Dr.R.Ganesan, Professor and Head of the Department of Botany,Annamalai University for his encouragement and help for allthe research works including the studies in ferns of Tamil Nadu.

The author could not forget the help rendered by Dr.K.Arumugam, the then professor and Head, DDE in Botany and now Registrar, Annamalai University and Dr.P.Sundaramurthy, the present professor and Head, DDE in Botany, Annamalai University for permitting him to go through the herbarium specimens of Indian Ferns from the Department. He is also thankful to Dr.D.Dhanavel and Dr.L.Mullainathan, Professors of Botany, Annamalai University for their continued help in his various research activities.

The author is thankful to the Director, French Institute, Pondicherry, Librarians, Botany Department Library and University, General Library, Annamalai University for their kind help in permitting him to go through important books and Journals on Indian Ferns. He expresses his gratitude to his village people atKo-Athanur, Virdhachalam Taluk, for collecting the ferns from the bore wells of 50 feet depth and to Mr.Kannan, Chidambaram for having helped him in the collection of ferns from the water canal of Neyveli, Cuddalore District.

Contents

Chapter 1

Introduction

In this part of the book, the plants present in the plain areas of Tamil Nadu, that is from mean sea-shore areas to 500 feet elevations, particularly the borders of Tamil Nadu and Kerala are studied. It is hoped that the ferns of the plains have not been studied so far so much as those of hill stations of Eastern and Western Ghats of Tamil Nadu. Beddome (1976) has made extensive studies of the ferns of higher elevations of Western Ghats. The plants of the plains, therefore have been studied by the author for the first time, particularly in the Northern Districts of Tamil Nadu. Manickam (1992) has studied the ferns of Tirunelveli, Kanyakumari and Madurai Districts.

At present, the wild and cultivated ferns from Kanyakumari District to Madras have been described in this book. A large number of wild and cultivated species of ferns have been studied from Nagercoil, Martandum and the neighbouring areas. The soil type and climatic conditions in these areas are more pleasing and identical to those of Kerala. Some of the species present in Kerala are also availablein these areas,such as Venkode and Keeripparai, on the way from Martandum to Thiruvananthapuram.

The next place of occurrences of ferns are Neyveli and its neighbouring areas near Virudhachalam Taluk of Cuddalore District.

Some of the plants were made into herbarium specimens. Some of them were photographed. The plant parts have been examined under 20 x lens, dissection microscope and compound research microscope at various magnifications and Camera lucida diagrams drawn. All the plants and plant parts were drawn by the author. It is hoped that the illustrations are more important than merely describing a particular plant.

The herbarium specimens have been kept at author's home laboratory at Mariappanagar, Annamalai nagar – 608 002, Chidambaram. The isotype of rare and new ferns will be deposited at Botanical Survey of India, Coimbatore, Tamil Nadu.

The author has presented papers on ferns of the plains of Tamil Nadu (Subramanian. D, 2007) and species diversity of *Adiantum* in Tamil Nadu (Subramanian, D., 2010) in national conferences and symposiums.

As Botanic Garden in-charge for over 25 years at Annamalai University, up to his retirement in 2002, the author had the experiences of the past 30 years to collect the ferns and other flowering plants from hill stations and various places of Tamil Nadu and grow them in the Botanic Garden for ornamental and research purposes.

As a staff member, the author had a large number of official and personal tour programmes to various hill stations and important Botanical Gardens throughout India and studied various groups of plants for the past 50 years.

The more important event was to see for the first time the few hundred diagrams of Ferns of Tamil Nadu in a book published by Beddome (1856) with descriptions of them also. The author happened to see this book with life illustrations of ferns in 2001 in a corner of Library of Botany Department, Annamalai University. Immediately, the author made xerox copy of the book, but a few pages became damaged. After a few months, when the author tried to see that book, the book got damaged totally and it was not seen afterwards. All the pages of the book became broken into pieces.

The author could identify the ferns of south Indian hill stations like Ooty, Kodaikanal, and Valparai only with the help of this book. Sir Beddome had collected all the ferns of Tamil Nadu, made herbariums and identified them by referring them with type herbarium specimens of International Herbarium Centre, at Kew, London and then with illustrations for all the ferns, he published them. As a Chief Conservator of Forests, Tamil Nadu, his outstanding contributions on Indian Ferns due to his self-interest and enthusiasm will be remembered by the scientific world forever.

Even though, Beddome had studied the ferns of Tamil Nadu, the ferns of the plains of this state have been almost ignored. Therefore, as a part I of ferns of Tamil Nadu, the author has studied the ferns of various plain areas of Tamil Nadu, from mean sea level to 500 feet elevations. The important places are Chidamabram, Virudhachalam, Cuddalore, Neyveli of Cuddalore District, Nagercoil, Marthandam, Keeripparai and Venkode of Kanyakumari District, Pondicherry, Madras, and neighbouring places of Coimbatore District.

Mere descriptions of plants and their parts will not be so much useful to understand or identify a particular plant as already stated as those of diagrams of them. It is more important in ferns and mosses than the other groups of plants. Therefore, the author has drawn almost all the plants and their parts as possible, in his publications in general and the ferns in particular.

The plain area of Tamil Nadu has the hottest summer and rainfall during November, December and January only every year. During summer, the maximum temperature goes up to 44°C. But the ferns are mostly cool loving plants. That is why, most of the plants are living at more than 5000 feet elevations from M.S.L. in hill stations of Western Ghats of Tamil Nadu.

Rarely or very rarely these ferns live in the plains of Tamil Nadu. Nobody cares about them. But they are considered as more important medicinal plants of the world. In the present investigation, most of the plants have been collected from the neighbouring places of Kanyakumari and Coimbatore in cool climatic conditions. A large number of ferns are being grown in nursery gardens with care for ornamental purposes in the plains of Tamil Nadu. The wild plants are very rare in those plain areas. So far, there is no satisfactory work on ferns of the plains of Tamil Nadu. So, present investigation has been undertaken.

The author is trying to grow rare and important ferns in his hosue garden at Mariappa Nagar, Annamalai Nagar – 608 002 and in his farm house garden at Ko-Athanur of Virudhachalam Taluk of Cuddalore District, his native place.

The Government and nature lovers should grow at least a few ferns in their house gardens or private gardens. Then, it is possible to save these rare and precious plants from extinction.

Chapter 2

Observations and Discussion

1. Ferns Collected from Virudhachalam

Ko.Athanur, Virudhachalam Taluk, Cuddalore District, the native place of the author is a plain area. In this village and the neighbouring villages, Mavidanthal, Karkudal, Kumaramangalam and Sathamangalam, there were bore wells of 50 feet depth and 10 feet diameter for setting motor pump sets for agricultural purposes.

Inside these wells, ferns like *Nephrolepis cordata* (Plate 27), *Pteris vittata* (Plate 24), and *Adiantum capillus veneris* (Plate 3) were present all over. With the help of village people, the author could collect these ferns some 20 years back and study them. Now, these bore wells have been avoided and people started to set up soil level bore wells of 7 or 8 inches diameter with plastic pipes. But even now, these ferns are available in old brick walls along with mosses in rainy seasons. *Cheilanthus albo-marginata* (Plate 31) is also present along with *Nephrolepis, Pteris* and *Adiantum.* Those plants are present only in the rainy reasons. But these ferns are present throughout the year in hill stations like Ooty Kodaikanal, Valparai, Kollimalai Hills, etc. When compared to the plant -, (Plate 34) collected from Ooty and Kodaikanal, the plants of this species collected in my village were, erect and rigid (Plate 37). Besides, there were differences in leaf characters and sori arrangement. Therefore, it was named herein as *Cheilanthus tenuifolia* var. *rigida* var. nov D.Subramanian (Plate 37).

The author had been growing these ferns in his Botanic Garden, Annamalai University, Chidambaram. They can be collected now-a-days in old brick walls of damaged and abandoned buildings during the rainy seasons.

2. Ferns collected at Chidambaram and Annamalai Nagar

In an abandoned building at Annamalai nagar, there were 2 ferns along with mosses and liverworts on a very old brick wall during rainy season. The tips of these ferns on touching the wall, the tips became swollen and seedlings were produced from the tips. The two walking ferns are varieties of Lindsaea repens. The first plant had leaves like those of Adiantum caudatum. Therefore, it was named as Lindsaea repens var caudatum var. nov. The second plant had almost spherical leaves and so it was named as Lindsaea repens var. rotundifolia var. nov. (Subramanian, 2007) (Plates 17).

All the species of *Lindsaea* so far reported from India, Srilanka and Malaya were compared with these two varieties of *Lindsaea*.

These two plants had no sori even though the plants were larger and matured.

There were *Polypodium phyllitides*, the "Strap fern"(Plate 38), *P. aureum glaucum*, *Adiantum trapeziforme* (Plate 11), *Nephrolepis duffii* (Plate 28), *Pteris cretica* (Plate 18), *P. cretica – albolineata* (Plate 19) - the "Variegated table fern", and *P. ensiformis –*victoriae (Plate 20) - the "silver table fern" available in the local nursery gardens at Chidambaram. My friend Dr.Murugaiyan, Professor of Linguistics at Annamalai University, had maintained a nursery plant garden at Mariappa Nagar of Annamalai Nagar and all the ferns mentioned above were grown in his garden, as ornamental and curiosity plants for sale. We have to take intensive care to grow them under shade and spray cold water intermittently particularly during summer hot days in our plain areas.

3. Plants collected at Kanyakumari, Nagercoil and Marthandum

In 2011, the author went to Nagercoil for 3 days and from there went to various places on the way from Nagercoil to Thiruvananthapuram, Kerala. Dr.N.Vijayakumar, Professor of Botany, S. T. Hindu College, Nagercoil helped the author in collecting Mosses and Ferns and some flowering plants. We collected them from Marthandum, Venkode and Keeripparai.

Acrostichum aureum is a large bushy shrub present as distinct colonies on the borders of water canals among the paddy fields. The fertile brown fronds are present at the tip of sterile shoots. These plants have been treated as one of the mangrove associates usually present in salt water areas and back water canals near sea-shores. But the presence of this plant in pure cultivated soil areas in Venkode on the way to Thiruvananthapuram is an interesting and unusual observation (Plate 50).

On the large borders of paddy fields and water canals at Venkode, there were large number of *Lycodium* ferns. There were 6 distinct types differing in vegetative leaves and fertile fronds collected and studied. They are distinct varieties of the two important species of *Lycodium*, namely *L. flexuosum Sw. (L.Semibipinnatum.R.Br.)* and *L. microphyllum* (Cav.) R. Br. (*L. scandens* (Linn) Sw.) ( ).

These species of *Lycodium* are usually present in higher attitudes of hill stations like Ooty, Kodaikanal and Valparai of Tamil Nadu. But, they are not present in the plain areas of most of the districts of Tamil Nadu. The climate of Kanyakumari areas are almost similar to Kerala. Therefore, *Lycodium* species are present in the plains of Kanyakumari of Tamil Nadu in the same way as these species are living in the plains of Kerala. Even though wild, these plants of *Lycodium* are more beautiful and peculiar than the ornamental plants (Plates). Now a days, due to climatic changes and human disturbances, these plants are fast disappearing. Therefore, we have to take suitable conservation methods to save these curiosity plants from extinction.

There was a walking fern, *Adiantum* species resembling *A.caudatum*on the rocky substratum of Venkode, Kanyakumari District. The plants were erect and tall and the pinnules were small and numerous (Plate 13). The tip of the frond on touching the ground produces a seedling like other walking ferns.

There was another plant resembling the previous fern, but smaller in size. It was another walking fern, and identified as *A. caudatum* from the marshy bogs of Marthandam (Plate). *Lindsaea ensiformis* (Plate 15), *Cheilanthus mysorensis* (Plate 33), and *Pteris vittata* (Plate 21) were also collected from Nagercoil areas.

There was a species of *Adiantum* resembling *A. lunulatum* but smaller in size withsmall pinnae (Plate 6). The fronds were erect and the pinnaewere beautifully arranged. It was collected from Keeripparai. A large number of characters are different from *A. lunulatum* and so this plant is named herein as *A. lunulatum* var. *minima var. nov.* The morphological characters will be described later. *Sphenomeris* species is alsoan ornamental fern collected from Marthandum (Plate 30). *Adiantum hispidulum* another walking fern rarely collected from Marthandum is also a wild fern (Plate 5).

There are more percentage of ferns following shoot tip propagation in the plains of Tamil Nadu. When compared to the hill station plants, *Adiantum* and *Lindsaea* species follow this method in the plains. The reason is that the species of the plains are lesser fertile, producing poor percentage of spores, which may be mostly sterile. Therefore the ferns in the plains follow vegetative propagation like seedling formations from the rhizome and shoot tip propagation. Spore germination into seedling may be avoided in some of the ferns in the unfavourable hot climatic conditions of the plains in which they are living. But when these ferns live in higher elevations of hill stations, they may produce more number of fertile spores and so sexual propagations may be followed.

Some of the ferns living in the plains and lower elevations of hill stations of Kerala are also present in various places of the plains of Kanyakumari District. It is because of the cooland favourable climatic conditions present in Kanyakumari district of Tamil Nadu like those of Kerala. Similar is the case in various places of Coimbatore District.

Adiantum heterophylla and *Ceratopteris thalictroides* are rarely present in swampy places of Kanyakumari District (Plate 56). The latter plants are comparatively smaller and shorter than those of Cuddalore District (Plate 55).

Many varieties of a fern collected at various places of Kanyakumari District like Nagercoil, Martandum, Venkode, Keeripparai and Thirunelveli, resemble *Microlepia pinnata* Moore, *M. platyphylla, Nephrodium propinquum* Br., and *Polypodium walkerae* Hook. with inverted V-shaped sori arrangement and general appearance. But, in detailed studies, the present plants are different. The present plants are distinct varieties of *Amphineuron terminans* (Hook) Holtt., (*Nephrodium terminans* Hook, *N. pteroides* Retz; *Polypodium* steroids Retz; *Cyclosorus interruptus* sensu. Holttauna) of the Thelypteridaceae as stated by Nayar and Geevarghese (1993) in their book "Fern flora of Malabar (Pages 300 to 302). They havedescribed this plant with illustrations. But, this fern has not been described by Beddome (1856) from Tamil Nadu. Very rarely this plant has been already collected from Valparai and Ooty and kept by the author unidentified. This plant has inverted 'V' shaped soral arrangement but the pinnae are distinct touching only by marginal lobes. In between the soral arrangement and mid vein, there is a large empty space on either side of the pinnae.

At Tirunelveli, Nagercoil and Venkode of Kanyakumari District, silver ferns are rarely available. They are looking like the species of *Cheilanthus*in general appearance. Sometimes, on the back side of the pinna of *Cheilanthus* species, there is white powdering like that of silver ferns.

Nayar and Geevarghese (1993) have described this silver fern in their book (*Pityrogramma calomelanos)* available in the plains and lower elevations of hill stations of Kerala. The author has already collected these silver ferns and gold ferns at Ooty, Kodaikanal and Valparai some 30 years back when he was a research scholar and then staff member of Annamalai University.Now several varieties of these two ferns were collected from these Hill stations by M.Sc., (DDE Botany) students and deposited in this Department, Annamalai University. A.B. Graf (1981) in her book "Exotica" has provided photographs of these silver ferns and gold ferns ornamentally grown in foreign countries. According to her, the silver fern is called *Pityrogamma calomelanos,* the gold fern, *P. chrysophylla* and California gold fern, *P. triangularis.* In 2011, the author has collected this silver fern along with varieties of *Amphineuron* and kept them, unidentified so far, in Kanyakumari areas in the plains.

Nayar and Geevarghese (1993) have described the Silver fern (*P. calomelanos* (Linn) Link) and Gold fern (*P. chrysophylla* (Sw.) Link) from Kerala. The former species is common, occurring in the plains and below 700 metres of hills (compoundwalls, walls of wells, on the soil, etc.) but the latter species is not common, rarely occurring in the plains but usually above 1000 metres of hills. According to Nayar, when both these ferns occur together in the same area, intermediate forms may be present. Excepting from Kanyakumari District, these silver ferns have been not reported up to last month from other districts (from the plains) of Tamil Nadu.

The other important point is that the author has fortunately drawn the silver and gold ferns collected from Kodaikanal, Ooty, and Valparai at elevations ranging from 6000 to 8500 feet. Both

Beddome and Nayar have not included the diagrams of these ferns in their books. Beddome has not at all mentioned these species from Tamil Nadu.

Kanyakumari and Coimbatore Districts of Tamil Nadu are almost identical in climatic conditions like that of Kerala. These districts are present near Kerala.

The ferns present in Kanyakumari, Coimbatore and Kerala in the plains are available only at 4000 to 6000 feet elevations of Ooty and Kodaikanal. The ferns available at 700 meters to 2000 meters of Hills at Kerala, will be available only from 6000 to 8500 feet altitudes of Ooty and Kodaikanal.

In the plains of Tamil Nadu, the ferns are present rarely. The wild ferns will be present only during rainy seasons in the plains of Tamil Nadu (excepting Kanyakumari and Coimbatore Districts).

4. Plants collected at Neyveli

Neyveli is present at the south western corner of Cuddalore District of Tamil Nadu. From the lignite mines of Neyveli, water is continuously pumped out and there is a continuous flow of water in large number of small and large canals. In a small water canal at Mantharakuppam of Neyveli, the author could collect some 7 types of ferns in the last two months (November and December, 2018). Of these, Acrostichum aureum is a larger fern, a small bushy shrub with fertile fronds at the tip of the aerial shoots. This plant is different when compared to the plants collected from, Venkode of Kanyakumari, Kozhikode, Kochi of Kerala and Goa plants (Plates 50 to 54). Nayar (1993) has stated that this fernis available only in salt water areas, particularly backwater canals of Kerala sea-shores. Now, the presence of A. aureum in the fresh water cultivated soil of Kanyakumari (Venkode) and fresh water canals of Neyveli which is far away, that is 60 km away from sea shore of Bay of Bengal is an unusual, observation.

The characters of *Acrostichum aureum* collected from various places of Tamil Nadu, Kerala and Goa will be described comparatively under the part 'Morphological Descriptions'.

Along with the above fern, the silver fern has been collected in the same water canal. When compared to the silver ferns of Kanyakumari, Ooty, Valparai and Kodaikkanal, the Neyveli Plant is different in some of the morphological characters (Plate 57). It may be due to the hot climatic conditions present in Neyveli when compared to Kanyakumari and hill station plants. Here, the gold fern is not available. The author will collect the silver and gold ferns from Kerala and compare them with the silver fern of Neyveli. This is the author's rare and unexpected observation. Now the author is growing all the ferns collected from Neyveli in his house garden at Mariappa nagar and itis successful. For the past 30 years, the author could not see this fern in other hot districts of Tamil Nadu. At Sooramangalam of Pondicherry, a variety of silver fern was collected by M.Sc. student and the herbarium specimen has been deposited at DDE, Botany, Annamalai University. The plant is shorter with smaller pinnae.

Another interesting observation is the presence of a large variety of *Pteris* (Plate 22). This is larger than *P. vittata*. By the side of Neyveli, the author has collected another variety of *Pteris* at Poondiankuppam Village of Cuddalore District on an old brick wall. This is shorter than the previous plant and has lesser number of pinnae. But the pinnae are longer than those of the previous plant (Plate 23). We have to identify these two varieties of *Pteris*. A variety of *Pteris* collected at Thalakonda of Chittoor district of Andhra Pradesh resembles the *Pteris* varieties collected from Neyveli and Poondiankuppam, but the Chittoor plant has broader and shorter pinnae with a basal bulging. At present, these three populations of *Pteris* may be considered as distinct varieties of *Pteris vittata*. The continuous flow of water through the canals, red-lime soil mixture and soil fertility are responsible for the growth and development of these ferns at Neyveli.

Along with the above said ferns, distinct types of ferns are presentin the same water canal. The first fern resembles *Nephrodium arbuscula* Desv. But the present plant is robust and rough to touch. The plant is smaller and small and large sori are found towards the margin of pinnae unlike the uniform sori present in *N. arbuscula*. Therefore, this plant is named herein as *Nephrodium arbuscula* var *robusta*var.nov. (Plate 47)

All the other 3 populations have inverted 'V' shaped uniform arrangement of sori towards the margin of the pinnae. The sori are restricted to the anterior free ultimate veinlets (so that sori are confined to the lobes of the pinnae leaving a broad sterile zone on either side of the midrib.). Therefore, they are treated as varieties of *Amphineuron terminans* (Hook) Holtt.of Thelypteridaceae. Even though the 3 varieties have the same pattern of soral arrangement, they are different in plant height, leaf shape, lobation of pinnae and general appearance. These varieties resemble closely to the varieties of *Amphineuron terminans* already collected from different places of Kanyakumari District (Plates 44 to 46). Nayar has described with illustrations onlyone plant from the plains of Kerala. But the author has so far collected several varieties of this species in Tamil Nadu. Now, it is confirmed that *Amphineuron terminans*is a fern capable of growing and establishing in the plains that too, in the hottest climatic conditions of Tamil Nadu like Neyveli, but so far the author has not seen this in other districts of Tamil Nadu, particularly in the plain areas.

5. **Plants collected at Coimbatore**

Climatic conditions are pleasant here. There are large number of small and large hills on the southern and western sides of Coimbatore. By the western side, the Nilgiri Hills started from Mettupalayam. On the southern side Pollachi, Udumalai Hills started and went up to Valparai of Annamalai Hills. On the north west, Kerala is present. These places have almost similar climatic conditions of Kanyakumari District. Therefore, a large number of ferns are available in the plains of Coimbatore. Important plants are Nephrolepis cordifolia (N. tuberosa), Lindsaea ensiformis (Plates 15, 25, 26), Pteris longifolia and species of Cheilanthus (Plates 32 to 34). Rarely the species of Adiantum are also available (Plates 1,2,4,7,8,12 and 14). Adiantum seemannii, A. macrophyllum, A. tenerum farleyense (Magnificent maiden hair), A. venustum, A.cuneatum, A. trapeziforme(Giant maiden hair), Nephrolepis biserrata (Boston fishtail fern) and N. cordifolia (Duffii, pygmy sword fern) have been ornamentally grown in private houses, nursery and gardens at Coimbatore. The author has grown these ferns in the Botanic Garden, Annamalai University (Plates 29, 30, 9,10).

Lindsaea ensiformis (Plate) is a beautiful fern resembling the species of *Pteris*. It was observed first at Venkode of Kanyakumari District (Plate 15). The same plant has been also collected from Palakkodu area (Tamil Nadu border), situated near Coimbatore (Plate 16). These plants are present in these plain areas. But in other plain areas of Tamil Nadu like Salem and Cuddalore, this species is not available. Therefore, the climatic conditions of Coimbatore and Kanyakumari are almost identical to those of Kerala. Even in the plain areas a large number of ferns are present in Kerala. The plants present at 6000 to 8000 feet elevations of hill stations of Tamil Nadu are available at 2000 to 2500 feet elevations of small hills at Kerala. The plants present at 3000 to 5000 feet elevations of hill stations at Tamil Nadu are available in the plains of Kerala.

Therefore, the climatic changes due to altitude in Tamil nadu is equal to the climatic changes due to combined effects of latitude and altitude in Kerala. Along with altitudes, the latitude in Kerala is responsible for more rain fall in a year and cooler climatic conditions.

Therefore a large number of ferns are available at Kerala borders of Kanyakumari and Coimbatore Districts, when compared to the other plain areas of Tamil Nadu.

As already stated, ferns are cool loving plants. Most of the ferns are living at higher attitudes of Ooty, Kodaikanal and Valparai.

They will be described in the second volume of book on ferns by the author shortly.

6. Plants collected at Madras and Pondicherry

Soundarya Nursery at Teynampet, Madras was an important nursery garden culturing more important ornamental plants particularly the ferns and orchids. Adiantum, Nephrolepis and Pteris with large number of cultivars are grown, propagated and sold there. Some 20 years back it was a pleasure to go to this nursery garden and purchase peculiar ferns, orchids and ornamental plants. Later on, it was shifted to Chengulput and the author has not gone there. Various varieties of Pteris cretica, Nephrolepis exaltata; and Adiantum tenerum, A caudatum, A. cuneatum and A trapeziforme (Plates) are sold there.

The Botanic Garden at Auroville near Pondicherry,Government Botanic Garden and Aurobindo Ashram at Pondicherry are more important places along with some private nursery gardens to grow ornamentalplants particularly ferns, balsams, roses, chrysanthemum, etc. They brought the hill station plants and established them in our climatic conditions. After a few years, they become fit to be grown in our houses and Departmental Botanic Gardens. They also imported exotic plants and in due course of time they became suitable to be grown in our area.

Each plant appears in this earth with a purpose and may be useful to human beings in one way or other. If we say a plant is useless, it means we do not know about that plant. If once a plant disappears from this earth, we cannot make it or create it at any means. It is the author's sincere request that we have tosave our plant wealth without losing it.

7) Plants collected at Courtallum and other places of Thirunelveli

1) Adiantum incism

In this fern, the leaves are peculiar and arranged on both sides of aerial fronds. The leavesare folded on the upper side of the lamina and arranged one above the other (Plate 7).

This fern is very rare and peculiar when compared to the type species described by Beddome and Nayar in higher elevations of Western Ghats and Kerala respectively. It was collected at the foot hills of Courtallum.

2. Cheilanthus

Near the foot of small hills of Courtallum, a small fern with 5 to 6 arial fronds was collected. It belongs to the genus *Cheilanthus* (Plate 36) and resembles *C. mysorensis*. But *C. mysorensis*is a taller plant with basal small leaved branches, middle larger and small leaved branches from a medium erect rachis at the tip. But in the present plant, such clear cut differences are not present and the number of lateral branches are lesser in number. This plant must be identified as it is not matching with any of the specis of south India so far described.

At the foot hills of Thiruvannamalai, one such plant was collected and kept in the DDE botany, Annamalai University as a herbarium specimen (Plate 35). Therefore, this plant of *Cheilanthus* may be also available at the foot of other south Indian hill stations also and in the plains as well.

3. Pityrogramma calomelanos

At Thirunelveli, a silver fern was collected and the herbarium specimen of this is available at DDE, Botany, Annamalai University. It is a smaller plant when compared to those of Neyveli, Sooramangalam, Nagercoil, Ooty and Kodaikkanal. Therefore, it is the 4th silver fern available in the plain areas of

Tamil Nadu. At Kamman District a population of *Pityrogramma calomelanos* was collected and a herbarium specimen of this plant was kept at DDE, Botany, Annamalai University. This plant is small and short, nearly ½ foot in height. The pinnae are broader and smaller, each with 1 or 2 teeth on the margin.

Chapter 3

Morphological Descriptions

1. *Acrostichum aureum* Linn. var. *acuminatum* var. nov. collected from Neyveli (Plate 51)

A large terrestrial fern with thick erect underground rhizome having persistent leaf bases. Stem 3 to 4 feet high, 7 to 10 mm thick, oblique,green when smooth and shining in the upper portion;rarely branched, grooved; Lamina shortly petioled, alternate surrounding the stem, elongate oblong in outline, 60 to 160 x 20 to 30 cm, 1-pinnate with rachis similar to stem; sterile pinnae, similar in some upper pinnae part of the sterile pinna converted into fertile pinna so that both basal sterile and upper fertile portions present in the same pinna, above which all the pinnae are fertile; the fertile pinnae 2 to 3or 4 to 7, go on reduced in size above and all fertile pinnae have acuminate, sharply ending tips. But the sterile pinnae are larger and almost uniform, longer with notched tip like that of "Riccia" plant (Abryophyte) smooth on both surfaces, leathery to touch. Venation pinnate, with prominent midrib; devoid of prominent lateral veins, but bearing many closely placed slender veinlets which are indistinct, profusely reticulate with narrow, areoleselongated oblique to midrib devoid of included veinlets and progressively shorter towards margin. Sporangia spread densely over the entire lower surface excepting the midrib, makinga dark reddish-brown mat over the surface and mixed with many peltate orange-brown paraphyses. Spores light brown, elongated spherical, 40 to 60 μ in size.

Present on the sides of perennial water canal at Neyveli, Cuddalore district.

2) *Pityrogramma calomelanos* (Linn.) Link.

Rhizome 3-4 cm long and 2 to 3 cm thick covered by pale brown paleae; Fronds erect with stem 13 to 40 cm long and 2 to 3 mm thick, shallowly grooved in the upper half, lamina erect, obovate 20 to 30 x 8 to 24 cm and covered by dense silvery farina on lower surface, rachis bearing several loosely placed alternate primary pinnae, facing upwards, secondary rachis bearing secondary pinnae, pinnatifed and acuminate in the apical region but pinnate below with many subopposite oralternate, sessile ultimate pinnae. The shape of primary and secondary pinnae vary in several varieties collected from several plain areasof Tamil Nadu. Sporangia with shorter stalks present in between midrib and margin of the lamina on both sides.

Cytological studies must be made to understand the evolutionary trends among the populations showing biodiversity and exact taxonomical status of a particular variety.

The presence of distinct populations from higher altitude of hillstations like Ooty and Kodaikanal to the plain areas up to the sea-shores of Tamil Nadu is an interesting observation (Plate 57 to 59). While the flowering plants present in various altitudes of hill stations up to the plains show various ploidy levels, the same nature of polyploidy may be present in this silver fern also.

3) *Pteris vittata* Linn (*Pteris longifolia* Auctt)

This is a terrestrial species occurring in the plains of Tamil Nadu and Kerala in dry habitats. According to Nayar and GeeVargheese

(1993), there are two forms, one markedly larger than the other. They called them small form and large form possibly representing tetraploid and hexaploid cytotypes. Now the author has collected a few types from his native place Ko-Athanur, Neyveli and Pondiyankuppam, Recently a herbarium specimen of Pteris deposited at DDE, Botany, Annamalai University collected from Chittoor of Andhra Pradesh has been studied by the author. It is in between Athanur and Neyveli plants in size. Therefore, this species is highly variable. The original diploid plants of this species may be present in cool climatic conditions of hill stations like Ooty and Kodaikkanal. The plants present in the plains may be higher polyploids like tetraploids, hexaploids, pentaploids and octoploids. The author will try to study the cytological characters of various distinct populations of this species. This species had been already reported from the plains of Rajasthan, Calcutta, etc.

Plants medium sized 1 to 2 feet in height with erect or suberect unbranched rhizome. Fronds erect or spreading "U" channeled in the upper side, basal pinnae smaller with cordate base and broad lamina, gradually becoming larger and in the middle larger pinnae, unipinnate, pinnae longer and narrower, ranging from 5 to 25 cm and 5 to 15 mm, according to the variety, facing sidewards, sori arranged in a linear pattern below the margin on either side. Pinnae gradually becoming smaller and shorter towards the tip. The ultimate pinnae are facing upward, longer and larger, (plates 18 - 24). Chaudhary et al (2006) have enumerated a plant of *Pteris*, *P. vittata* Linn. from Rajasthan resembling the plants of this species from Ko-Athanur of Cuddalore district. The populations studied from different places differ among themselves in size of the plants, size of pinnules and number and arrangement of pinnules. But it is a highly variable species present very rarely in the hot climatic conditions in the plains of Tamil Nadu (Refer appendix, Table 1).

4) *Amphineuron terminans* (Hook) Holtt
(Plates 47 to 49)

A medium sized terrestrial fern with cylindrical creeping branched rhizome under the soil, fronds loosely placed, spirally arranged, stem erect, 40 to 80 cm long, 5 to 8 mm thick, tetragonal; Lamina ovate–deltoid, 1-pinnate, lateral pinnae 10 to 25 pairs, behind pinnatifid apical short region, the anterior half resembling lateral pinnae, but basal half broader and deeply pinnatifid with lobes resembling small lateral pinnae, venation pinnate, lateral veinlets simple and oblique. The sori are confined within the lobes of the pinnae, showing inverted 'V' shape. Between the sori and midrib, there is enough empty space. The varieties of this species collected from Neyveli and Kanyakumari are listed below.

4a. *Nephrodium arbuscula*. Desv Var. *robusta*. Var. Nov. D.Subramanian.

Of the four plants, the first plant (Plate) collected with Achrostichum at Neyveli resemble Nephrodium arbuscula Desv. But the present plant is robust and rough to touch. The plant is smaller and small and larger sori are found scattered under the pinnae unlike the uniform sori present in N. arbuscula. Therefore, this plant is named herein as Nephrodium arbuscula Desv. var. robusta var. nov. (Plate 47).

All the other 3 plants have inverted 'V' shaped sori restricted to the anterior free ultimate veinlets (so sori are confined to the lobes of the pinnae, leaving a broad sterile zone on either side of the midrib). Therefore, they are the members of *Amphineuron terminans* (Hook) Holtt, of The lypteridaceae. Even though these 3 types of ferns have the same sori arrangement, they are totally different among themselves in plant height, leaf morphology and general appearance. Nayar and Geevarghese (1993) have described only one species from Kerala under *Amphineuron*, that is *A. terminans*, But along with the 3 types collected at Neyveli, the author has collected 3 more types of plants resembling *Amphineuron terminans* from Venkode, Marthandom and Keeripparai of Kanyakumari District. All these 6 plants have inverted 'V' shaped sori restricted to lobes of the pinnae. These plants are having so much distinct characters as to keep them as separate species. But, the author likes to keep them as distinct varieties of *Amphineuron terminans* (Hook.) Holtt at present (Plates). Further works are necessary to come to abetter conclusion. The sevarieties were included under *Nephrodium* and *Microlepia* according to Beddome but Nayar put this inverted 'V' shaped sori bearing plant as *Amphineuron terminans* (Hook) Holtt (*Nephrodium terminans* Hook; *Nephrodium pteroides* Retz; *Polypodium steroids* Retz; *Cyclosorus interruptus* sensu Holttum) (pp.300 to 302) Nayar has put only one plant with inverted 'V' sori. But the author has now 7 plants with such soral arrangement.

The presence of these types of ferns at Neyveli and different places of Kanyakumari District is the first observation. The Cuddalore district having Neyveli and Kanyakumari District are far away from each other but the same types offerns presentin these two places is an interesting observation.

4b. Varieties of *Amphineuron terminans*

I. Plants from Neyveli

1. This first variety is larger with weak fronds having longer but narrow leaves in sub-opposite arrangement. The lobes more than ½ way are facing upwards with 'V' shaped inverted sori arrangement restricted to the lobes of the leaves (Plate 47).

2. This second variety is also larger but with rigid leaves. Leaves narrow and longer and the lobes face side ward. The inverted 'V' shaped sori are restricted within the lobes (Plate 48).

3. The third variety has medium sized fronds in opposite pattern. The lobes of leaves are facing upwards. Here also the sori are restricted within the lobes (plate 45).

II. Plants from Kanyakumari

4. The first variety is the largest of all varieties and resembles the type plant of *Amphineuron terminans* described by Nayar and GeeVargheese (Plate 46).

5. The second variety is also larger with longer and broader pinnae but slightly flexible. The lobes of pinnae are facing upwards with inverted 'V' shaped sori arrangement restricted within the lobes of the pinnae (plate).

6. The third variety has shorter and smaller fronds and smaller in size. The lobes of leaves face upwards with inverted 'V' shaped sori arrangement. The leaves are thicker and rough in texture (plate).

7. The fourth variety has the fronds medium sized. It almost resembles the first variety but leaf shape and lobation of leaves are different. The inverted 'V' shaped sori making a thick and continuous zig zag arrangement on the back side.

5. *Cheilanthus*

Beddome (1976) has described Cheilanthus fragrans Swartz, C. szovitzii Fisch and Meyer, C. mysorensis Wallich, C. fragilis Hook, C. varians Wall, C. laxa Moore, C. tenuifolia Sw, C. farinosae Kaulf, C. subvillosa Hook, E. albomarginata Clarke, C. rufa Don, and C. argentea Kunze, from India including Ceylon and Himalayan regions. But Nayar and Geevargheese (1993) have described only 3 species from Kerala namely Cheilanthus farinose, C. mysorensis, and C. tenuifolia. While most of the fernsare available in Western Ghats of Tamil Nadu only 3 species are present in Kerala. Some of the species of Cheilanthus present in Tamil Naduare not present in Kerala. As already stated, the climatic conditionin 2000 to 3000 feet altitudes in Kerala is equal to those of hill stations with 7000 to 8500 feet altitudes in Tamil Nadu. That is why even though such tall hill stations are not present in Kerala like Tamil Nadu, all the ferns available in Western Ghats of Tamil Nadu are available in Kerala.

The plains of Tamil Nadu except Kanyakumari and Coimbatore Districts, are drier with hot climatic conditions. Therefore the ferns are very rare and if present, they are completely different from those of hill stations of Tamil nadu and Kerala. Therefore, the populations of *Cheilanthus* at Courtallum and Thiruvanamalai are peculiar. They are not described by Beddome and Nayar. These populations resemble in the nature of pinnae to some extent to *C. mysorensis* but differ in size of the plants and general appearance. *C. mysorensis* willnot grow in such a hot climatic conditions such as Thiruvannamalai, Madurai, Tiruchi, etc. Therefore, the distinguishing characters of the present populations and *C. mysorensis* have been presented in a table (Refer appendix, Table 2).

Alongwith *Cheilanthusmysorensis, C. tenuifolius* isalsooccurring in the plains of Kanyakumari and Coimabtore Districts. Sometimes, the latter species may be also present in the lower elevations of Hill stations below 2000 feet altitude. Nayar (1993) has stated that these two species are present in the plain areas of Kerala. At present, the varieties of *Cheilanthus* resembling *C.mysorensis* have been collected from foot hills of Madras, Courtallum, Thiruvannamalai and Nagercoil and the illustrations are provided in this book. In general morphology and morphological characters ofplant parts, these varieties are different from *C.mysorensis*. Beddome or Nayar have not at all described these plants from the plains of Tamil Nadu or Kerala respectively. Anyhow, the author is treating them at present as distinct varieties of *C. mysorensis*. Choudhary and his co-workers (2006) have published a paper on bryophytes and pteridophytes of Rajasthan. They described *Cheilanthus albo-margin at a* with illustration occurring in the plains of Rajasthan. This species has been collected by the author from Maruthamalai foot hills of Coimbatore and foot hills of Kalrayan hills of Tamil Nadu with elevations of 500 feet from M.S.L. This species has been also collected from the plains of Kanyakumari District, Again, the presence of this species is a first observation from South India (Plates).

It is the observation of the author that the following ferns can beliving in the plains of Tamil Nadu.

1. **Ceratopteris thalictroides**

2. *Nephrolepis tuberose, N. exaltata* **and ornamental species of** *Nephrolepis*

3. *Pteris vittata* **varieties and ornamental species of** *Pteris*

4. *Adiantum capillus –* **veneris** *A. lunulatum* **and ornamental species of** *Adiantum*

5. *Lindsaea repens* **and** *L. ensifolia*

6. *Lycodium flexuosum* **and** *L. microphyllum,* **varieties**

7. *Acrostichum aureum* **varieties**

8. *Nephrodium flexuosum*

9. *Amphineuron terminans* **varieties**

10. *Cheilanthus mysorensis, C. albo marginata and C. tenuissima* **varieties.**

11. *Pityrogramma calomelanos* **varieties (Silverferns)**

12. *Polypodium* **and** *Cyrtomium* **species as ornamentally grown plants.**

For the ornamental ferns, great care is necessary to grow them in the plains of Tamil Nadu.

Conclusion

The ferns Acrostichum aureum, Pteris vittata varieties, silver fern, Pityrogramma calomelanos, varieties of Amphineuron terminans and Ceratopteris thalictroides available at Neyveli and its neighbouring places of Cuddalore District are also available at Nagercoil and neighboring areas of Kanyakumari District. The soil type and climatic conditions of these two places are completely different. Both of them are plains and far away from each other. It is interesting to note that so far the author has not collected these ferns in the vast areas in between Kanyakumari and Cuddalore Districts, particularly in the plains. Further attempts will be made to find out whether these ferns could be available in other Districts like Ramanathapuram, Thirunelveli, Madurai, Tiruchi, Tindukkal, Nagappattinam and Villupuram Districts of Tamil Nadu. In these places, the other ferns like Nephrolepis, Adiantum and Cheilanthus species may be rarely available. In general, the occurrence of ferns in the hot climatic conditions of the plain areas of Tamil Nadu is very rare particularly the wild plants. Some of the ferns have been grown for ornamental purposes. Even though the ferns are medicinally important nowadays, there is not so much realization of these purposes, among the common people. The wild ferns are facing more unfavourable situations for further existence. They are mostly endangered plants. Beddome has concentrated his studies only on the ferns of Hill stations like Ooty, Kodaikanal and Valparai. Therefore, it is perhaps, the first attempt, to study the ferns of the plains of Tamil Nadu. It is aimed in the present study to highlight the important ferns available in these regions and the need to conserve them without disappearing from this earth. The ferns are the most important, beautiful, and precious plants of this earth. The Government, the scientists and others who are interested in growing rare and important plants for various purposes must take care to establish these kinds of rare plants in our country. Following the publication of this first part of study of ferns from Tamil Nadu plains the next part on ferns of hill stations from 1000 to 8500 feet altitudes of Tamil Nadu will be published shortly.

Chapter 4

List of Plates and Figures

List of Photographs

Plate - 75   *Pityrogramma calomelanos* = silver fern. From Neyveli, upperside

Plate – 76   *Pityrogramma calomelanos* from Neyveli. Silver fern, lower side

Plates

Plate – 1   Fig.1 = *Adiantum radianum,* Fig. 2 = *A. venustum*

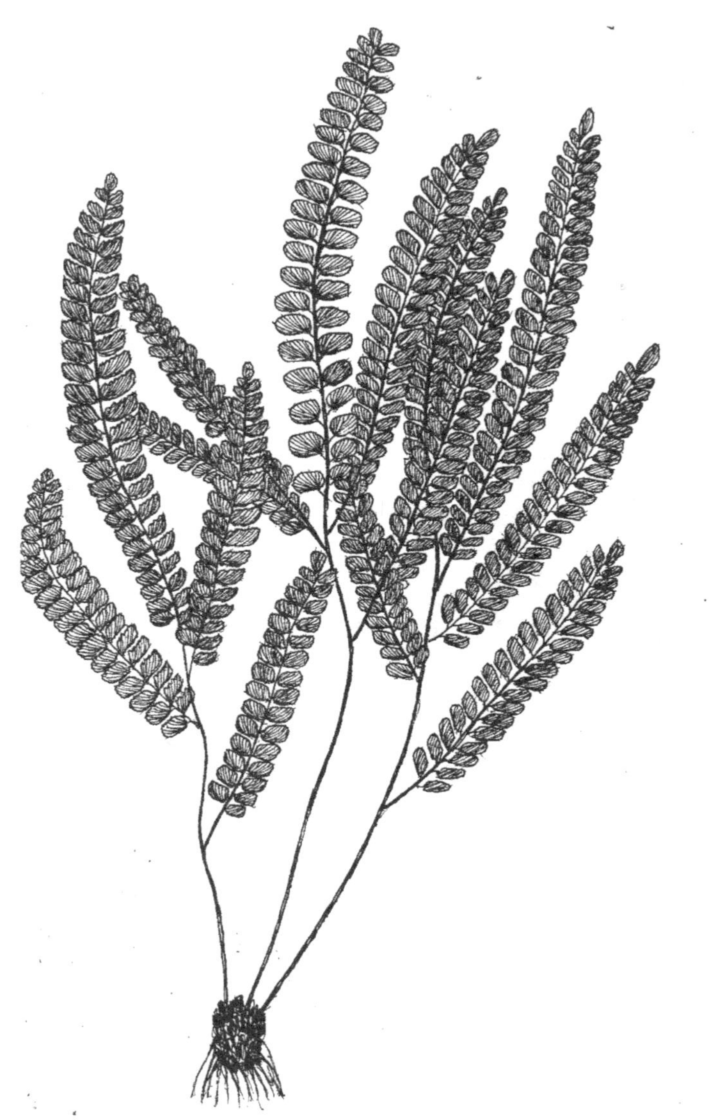

Plate – 2 *Adiantum hispidulum*

Plate – 3 *Adiantum capillus* – veneris

Plate – 4 *Adiantum seemannii*

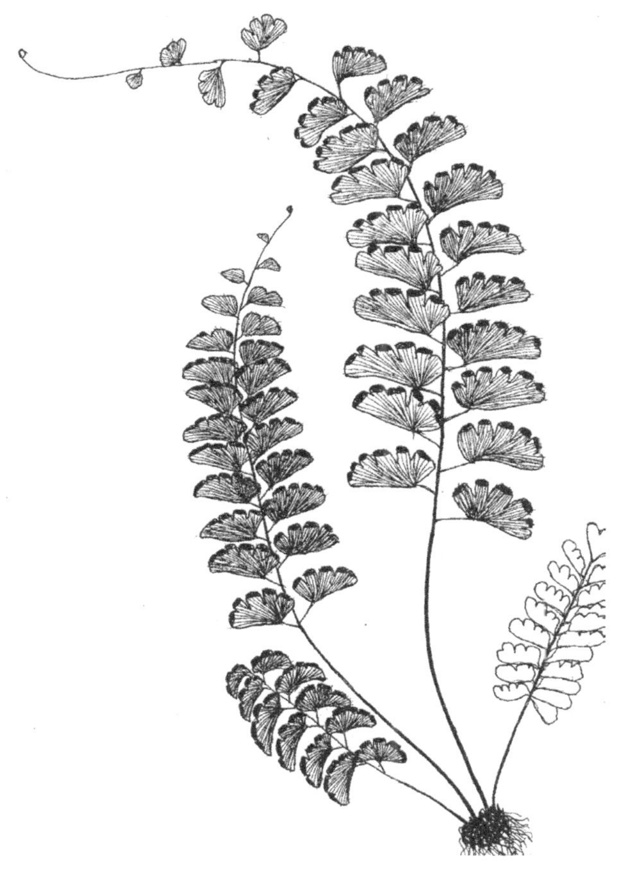

Plate – 5 *Adiantum hispidulum*

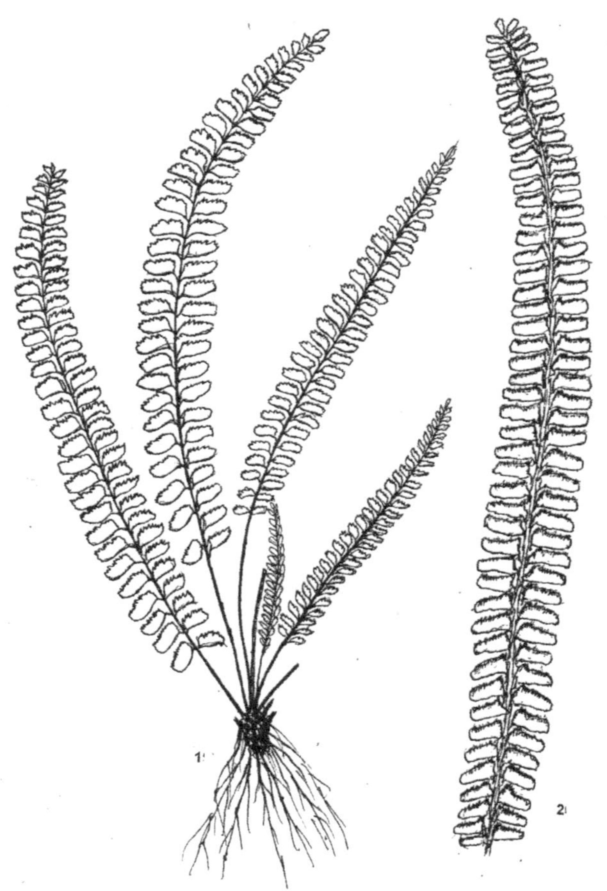

Plate – 6 Fig.1 = *Adiantum lunulatum*, var. *minima*, Fig.2 = *Adiantum lunulatum* var. *elongatum*

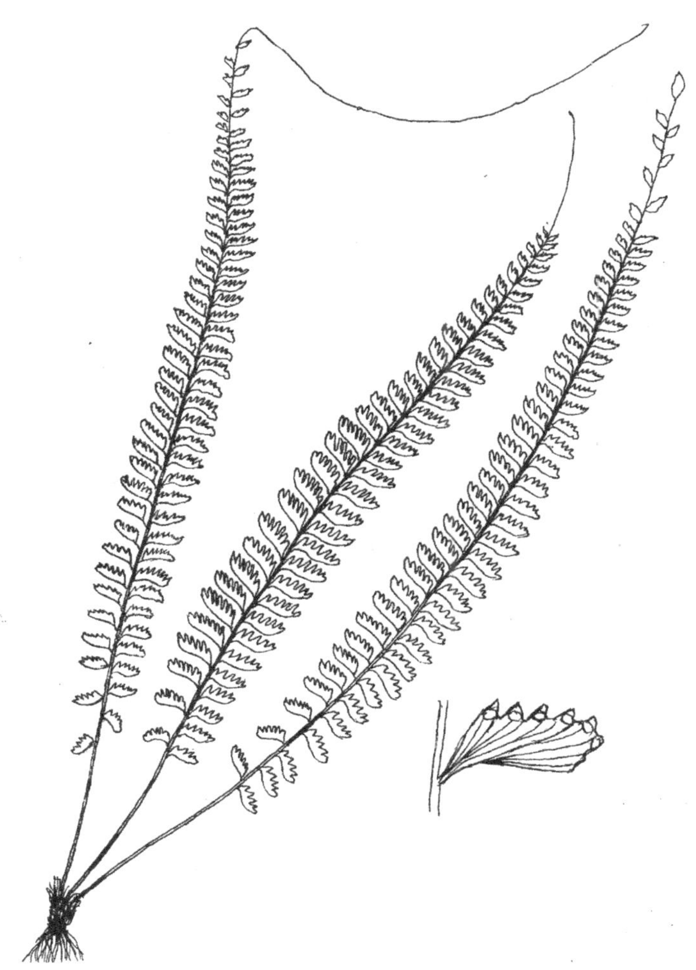

Plate - 7 *Adiantum incism*

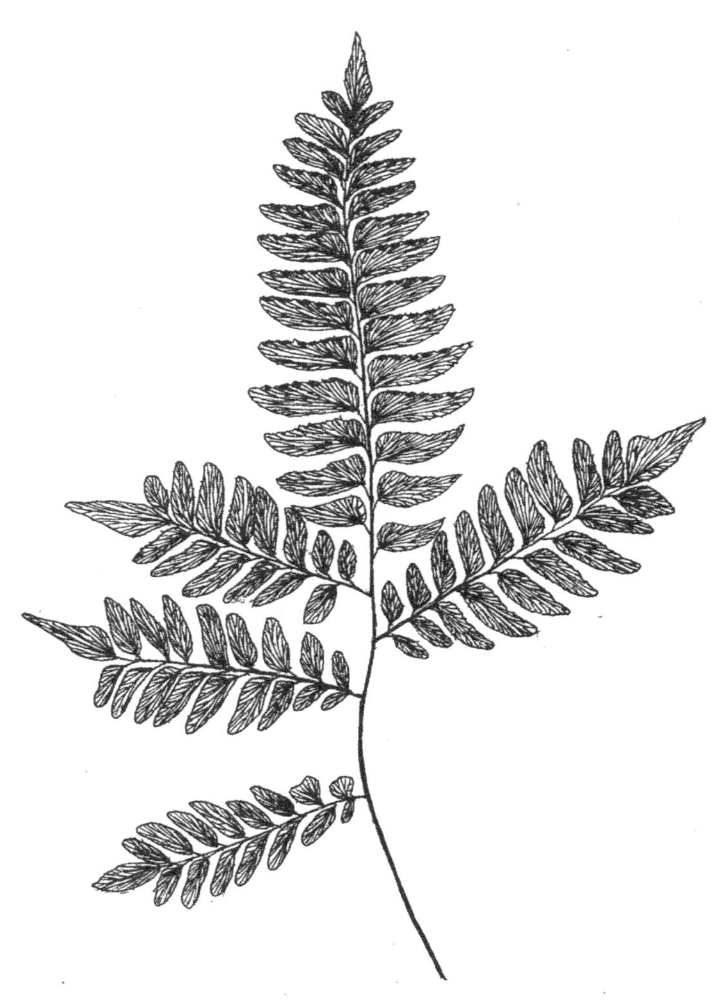

Plate – 8 *Adiantum*species

Plate – 9 *Adiantum tenerum*

Plate -10 *Adiantum tenerum*

Plate – 11 *Adiantum trapeziforme*

Plate – 12 *Adiantum macrophyllum* varieties

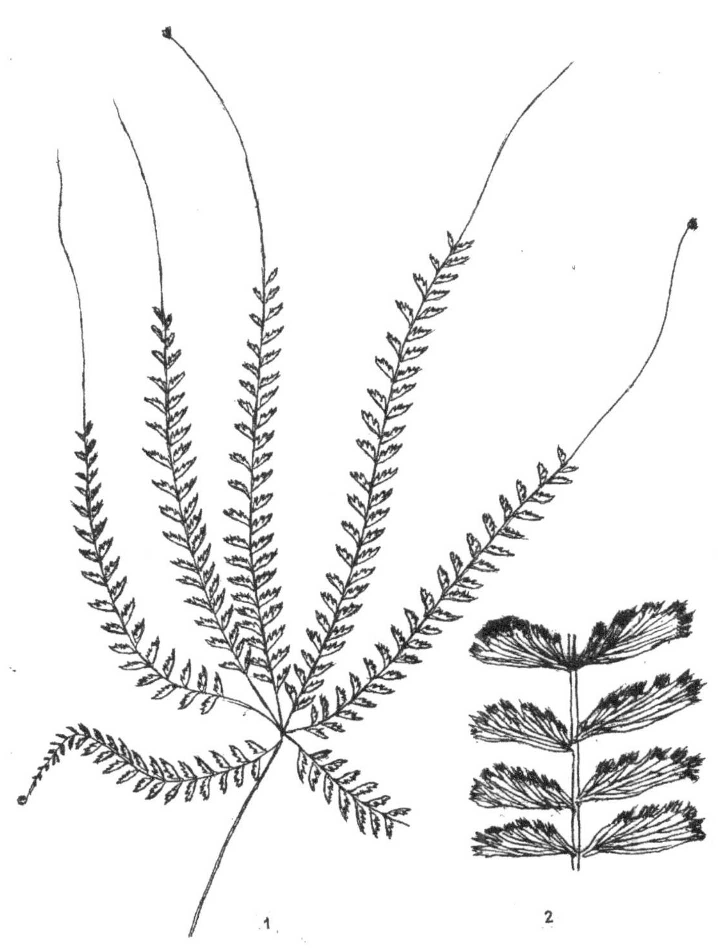

Plate – 13 *Adiantum caudatum*

Plate – 14 *Lindsaea heterophylla* varieties

Plate – 15 *Lindsaea ensiformis* kanyakumari type

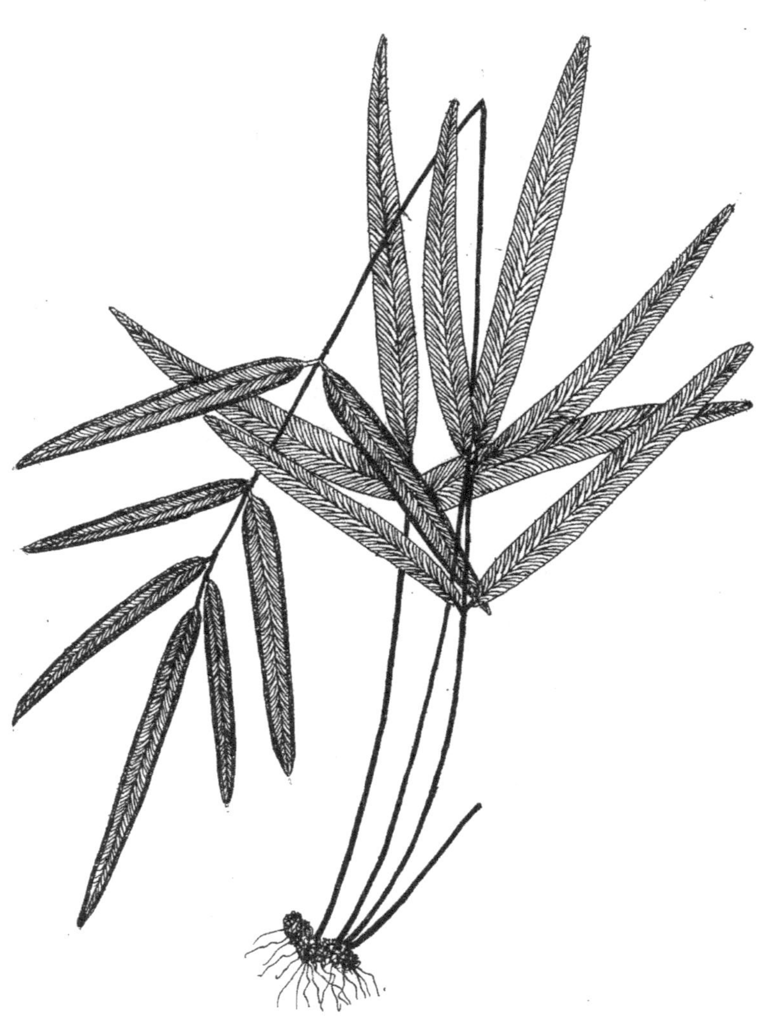

Plate – 16 *Lindsaea ensiformis* Palakkadu type

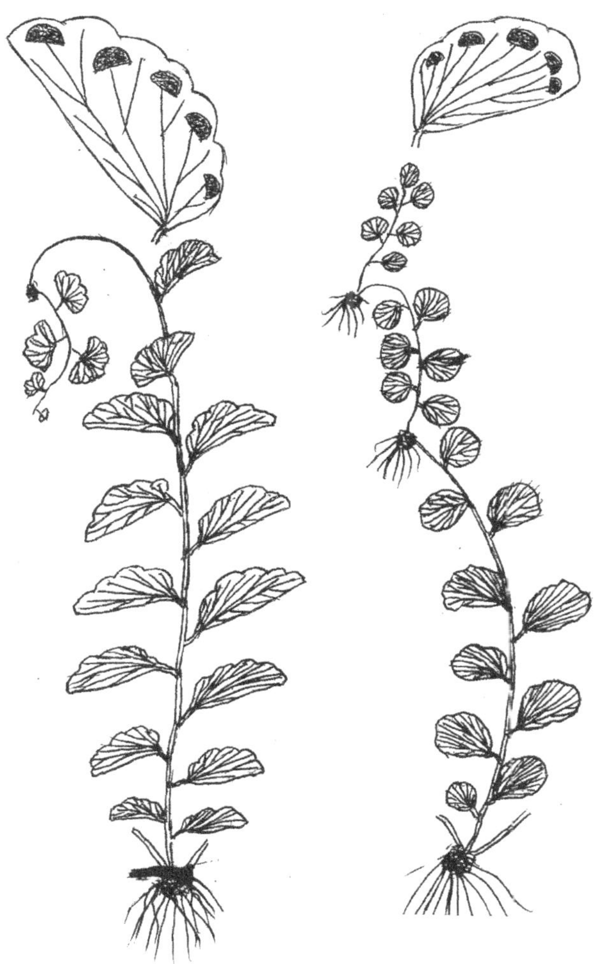

Plate – 17 **Fig.1** *Lindsaea repens.* **var.** *minor,* Fig.2. *Lindsaea repens.* var. *rotundifolia*

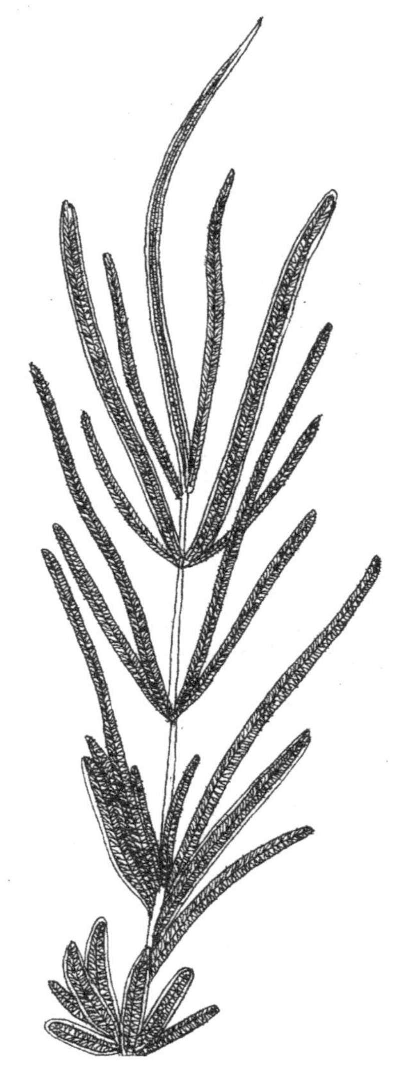

Plate –18 *Pteris cretica*

Plate – 19 *Pteris cretica albo-lineata*

Plate – 20 *Pteris ensiformis* **victoriae**

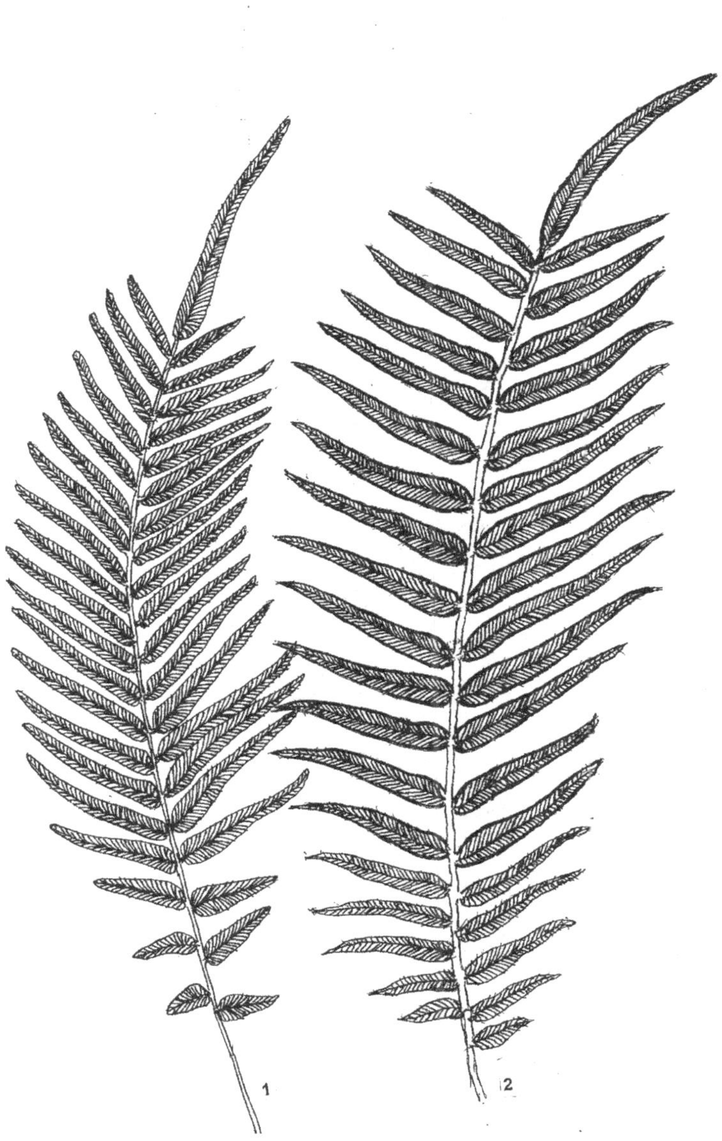

Plate – 21   *Pteris vittata* from Vencode, Kanyakumari

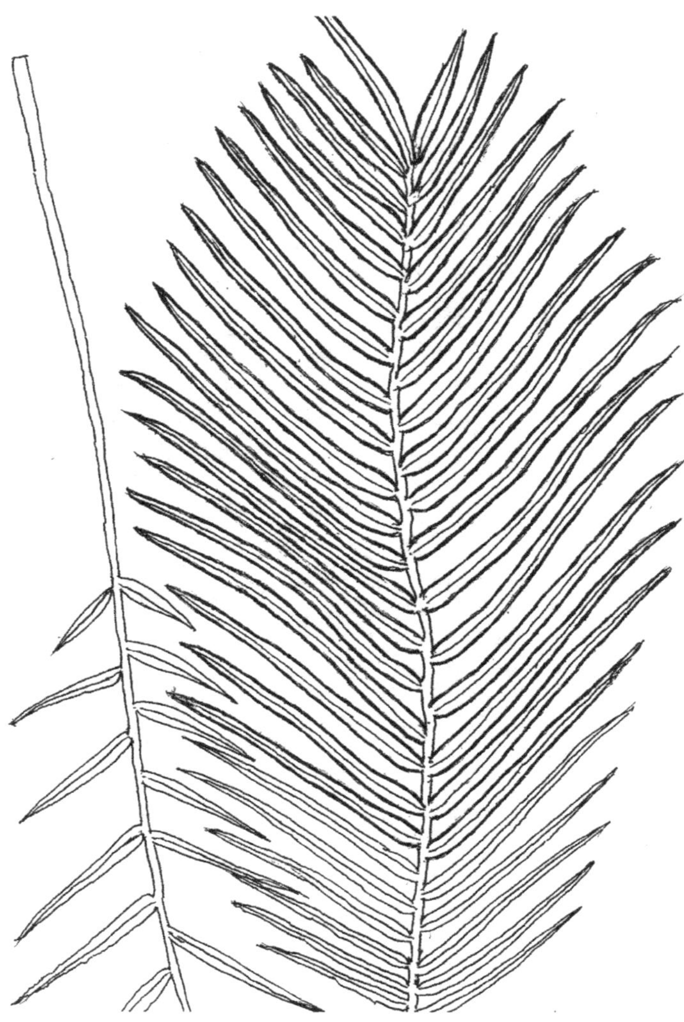

Plate – 22 *Pteris vittata* from Neyveli

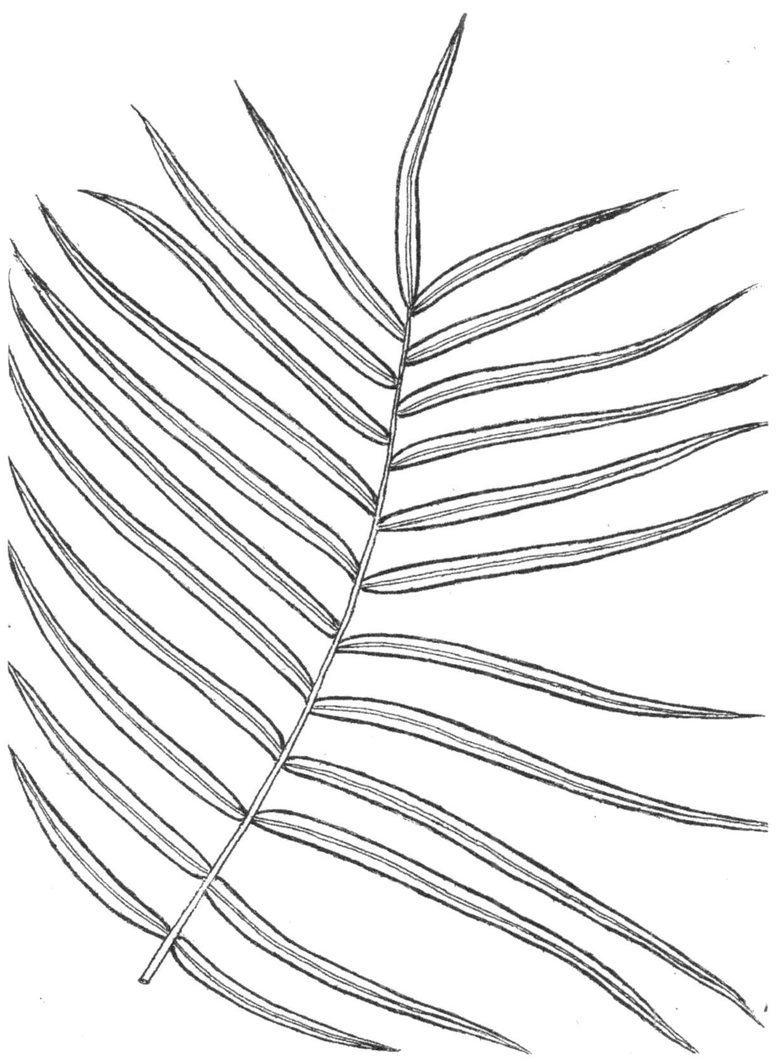

Plate – 23   *Preris vittata* from Poondiankuppam

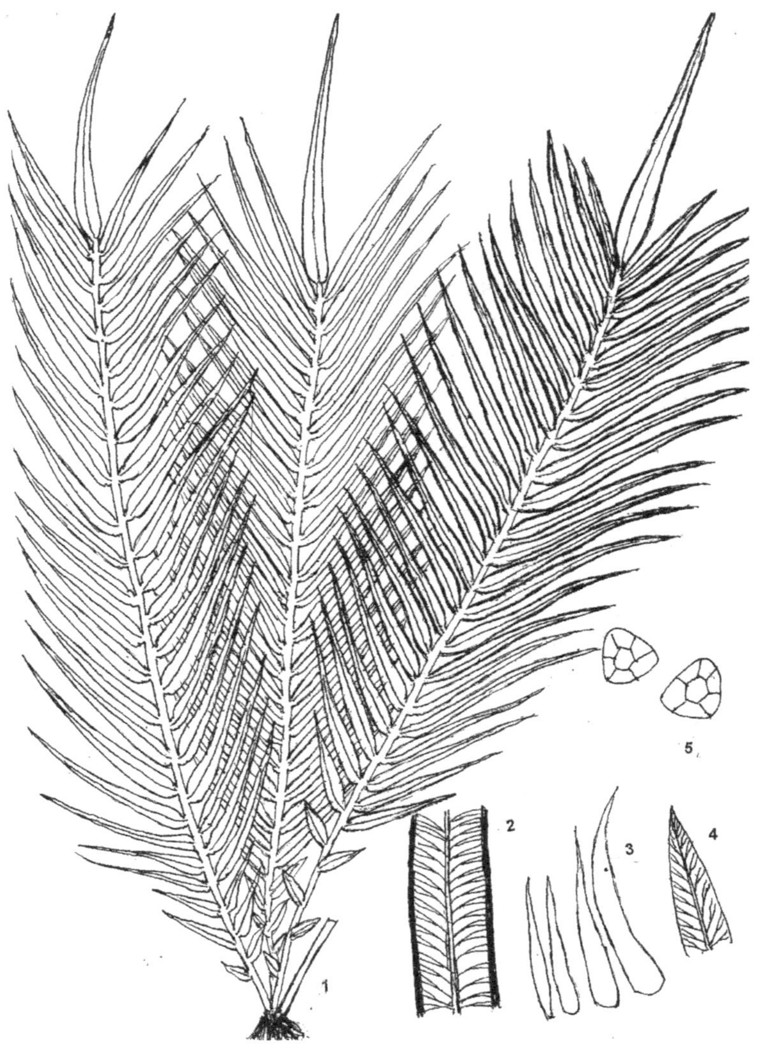

Plate – 24 *Pteris vittata* from Ko-Athanur

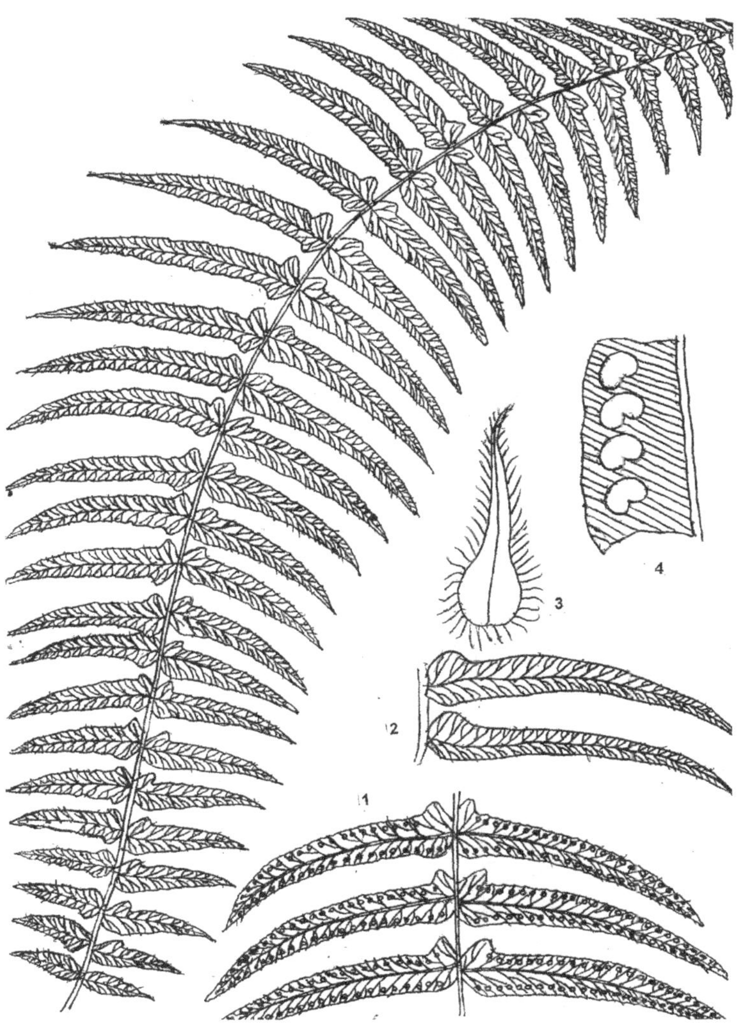

Plate – 25 *Nephrolepis exaltata*, Fig. 1 & 2 = Fertile and sterile pinnae, Fig. 3 = Scale,4 = Sporangium

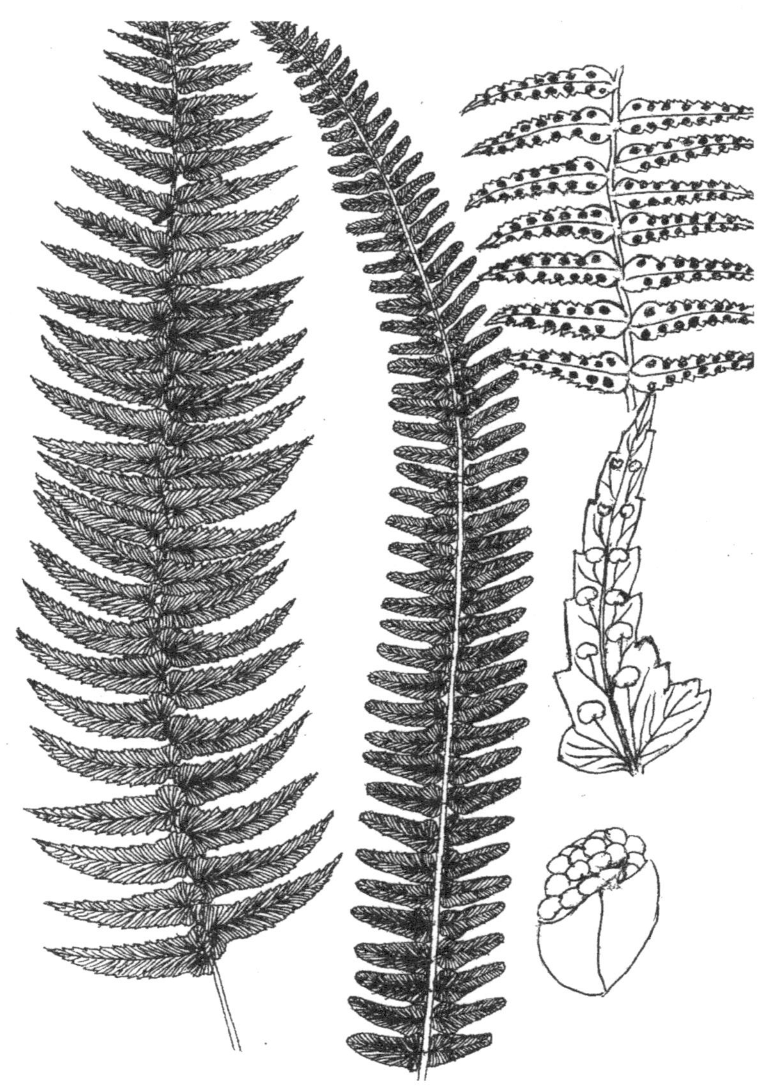

Plate – 26 *Nephrolepis hirsutula*cultivars

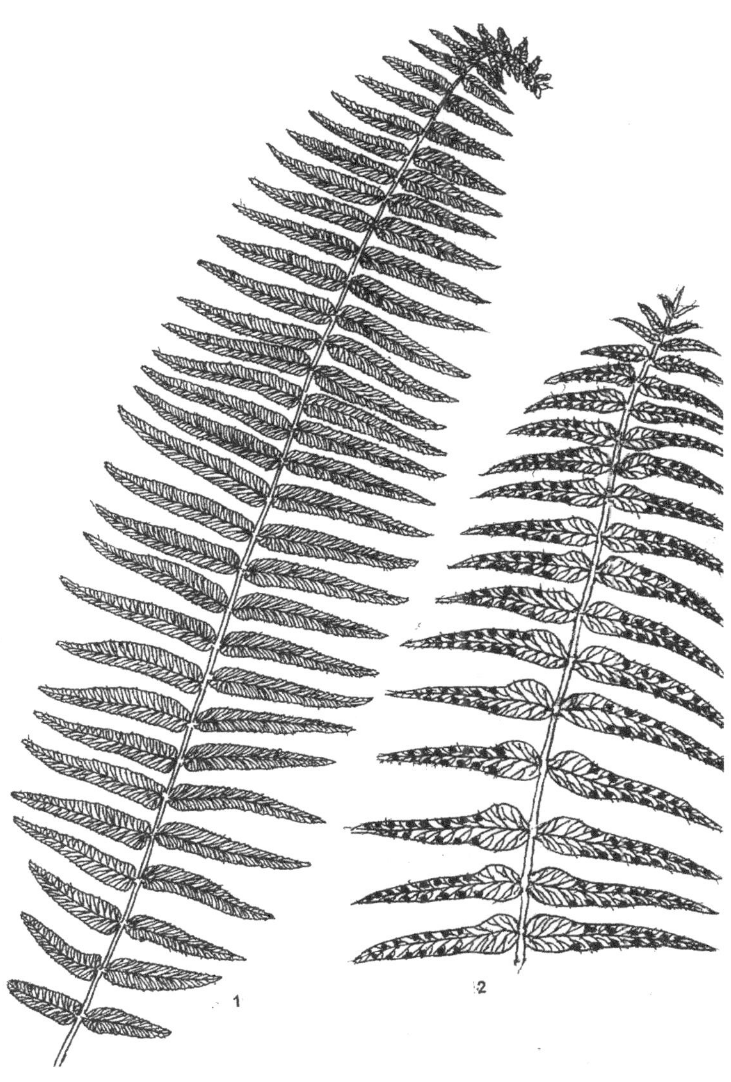

Plate – 27 *Nephrolepis tuberosa*

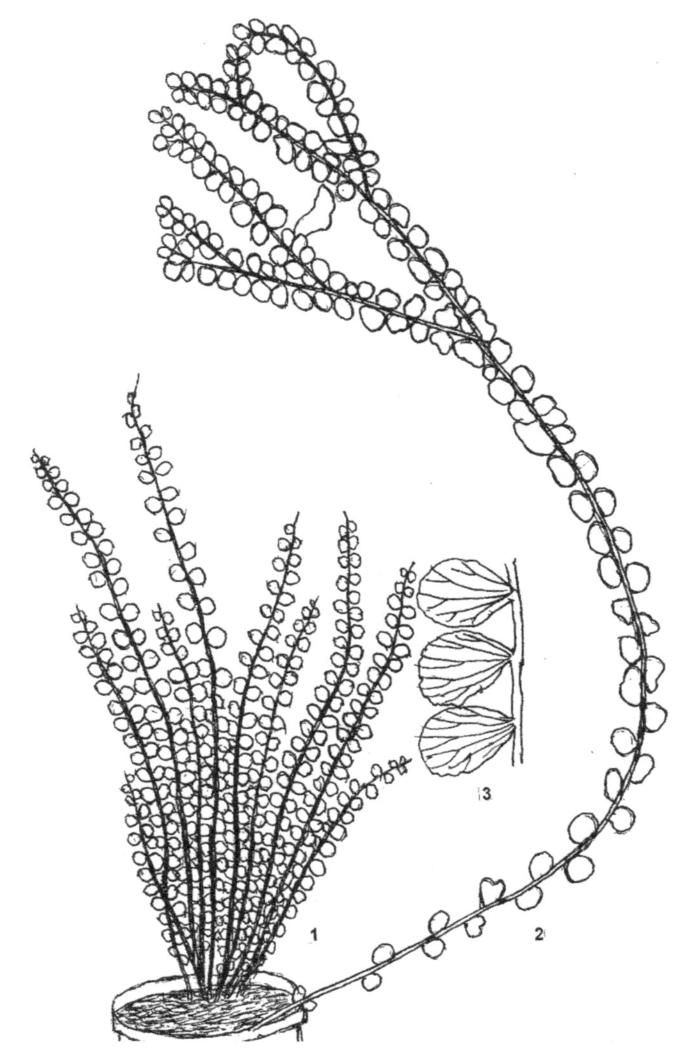

Plate – 28 *Nephrolepis duffii*

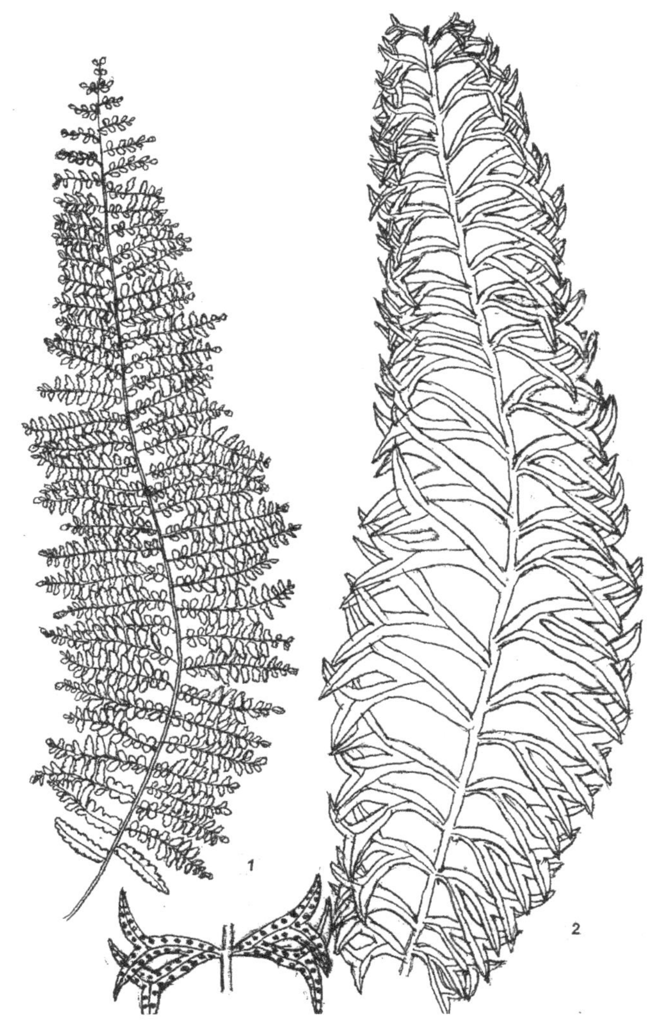

Plate – 29 Fig-1: *Nephrolepis exaltata,* Fig.2 *Nephrolepsis biserrata*

Plate – 30   *Sphenomeris* sp.Cultivated

Plate – 31 *Cheilanthus albo-marginata*

Plate – 32 *Cheilanthus albo-marginata* Kanyakumari varieties

Plate – 33   Figs 1 and 2 – *Cheilantusmysorensis*

Plate – 34 *Cheilanthus tenuifolia*, Fig. 2 and 3 = Sterile and fertile pinnae

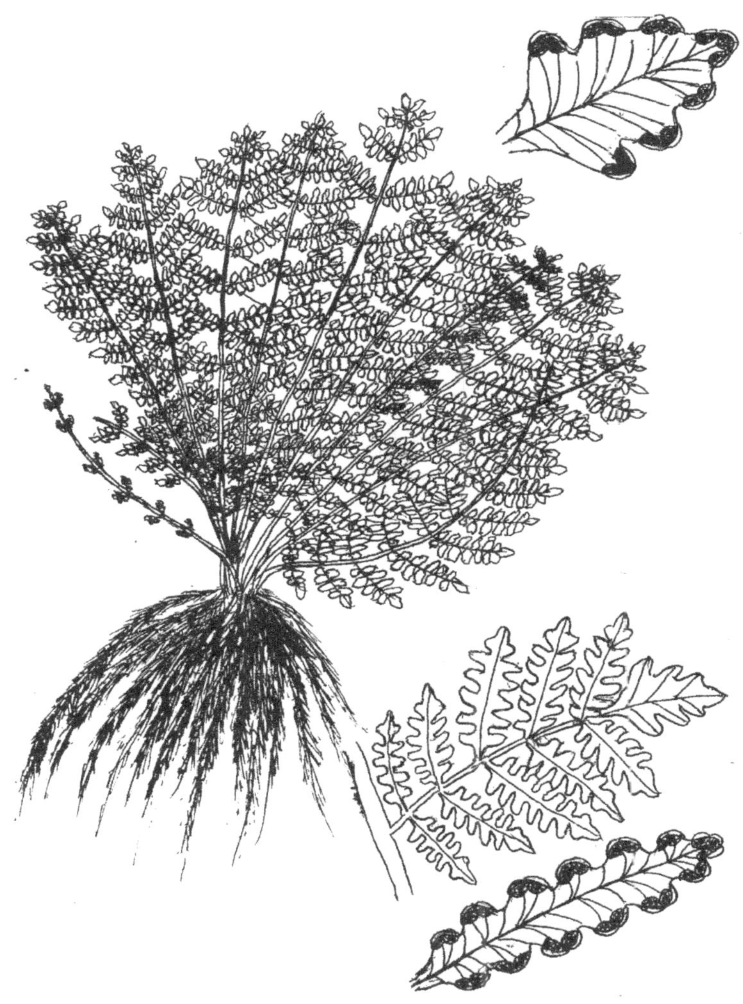

Plate – 35 *Cheilanthus mysorensis* Thiruvannamalaivariety, Fig. 1 – Plant, Fig. 2 - Sterile pinna, Fig. 3 - Lateral fertile pinnule and Fig. 4 - Tip fertile pinnule

Plate – 36 *Cheilanthus mysorensis* Courtallum, variety, Fig. 1 = Plant; 2 = Sterile pinna, 3 = Tip and lateral, fertile pinna

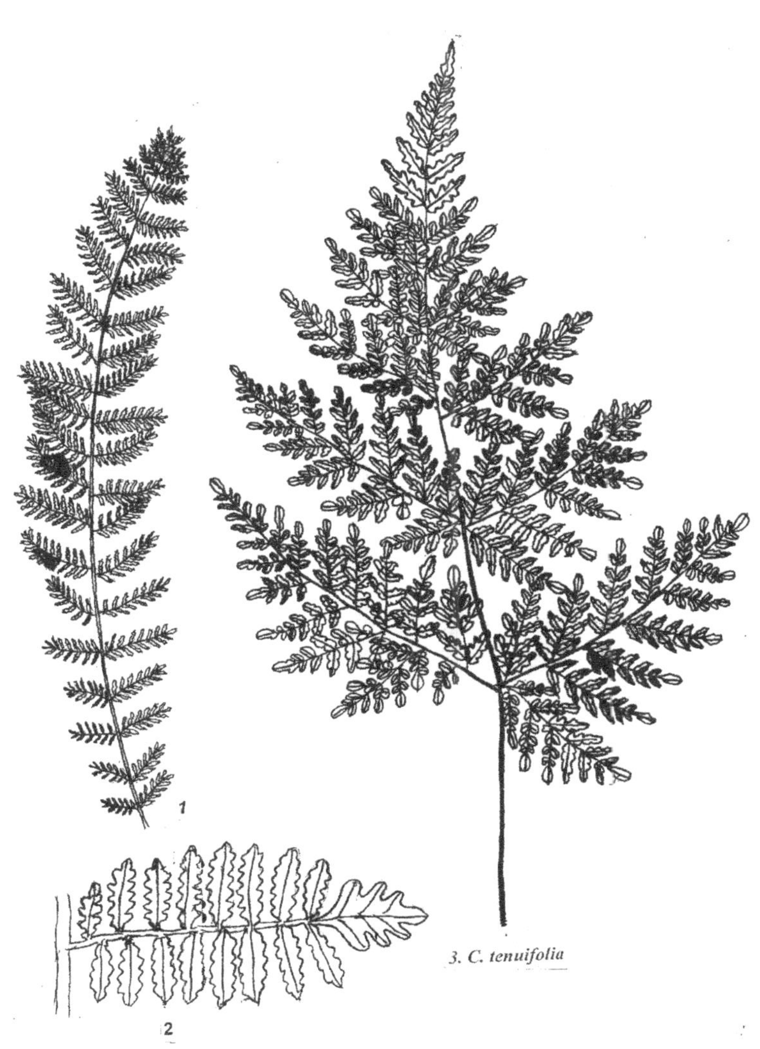

Plate – 37 *Cheilanthus mysorensis*

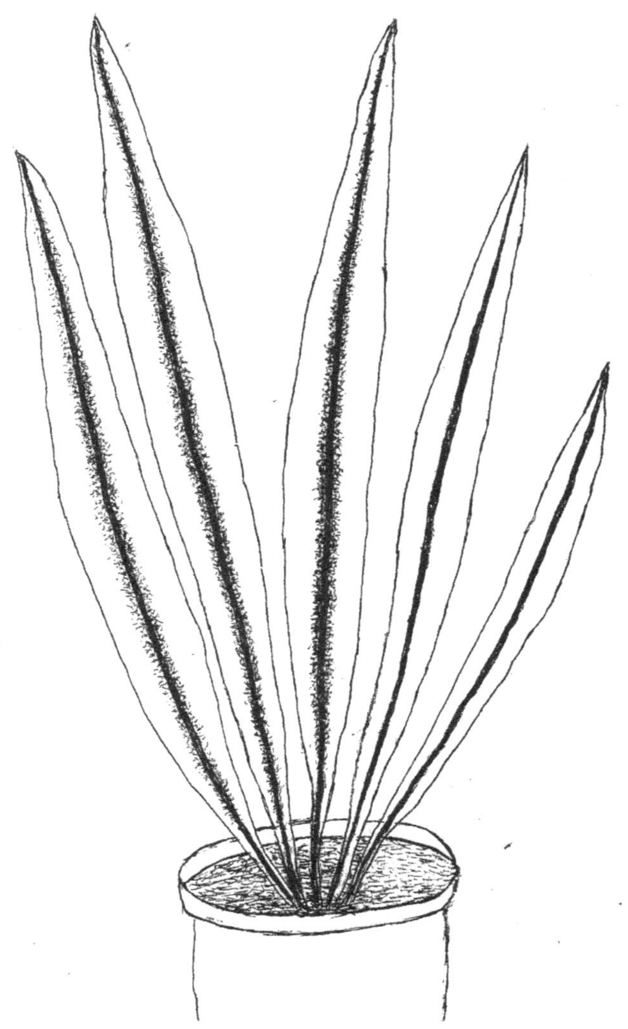

Plate – 38 *Polypodium phyllitides* ornamental form

Plate – 39 *Lycodium microphyllum*

Plate – 40 *Lycodium microphyllum*

Plate – 41 *Lycodium flexuosum*

Plate – 42 *Lycodium flexuosum*

Plate – 43   *Lycodium flexuosum*

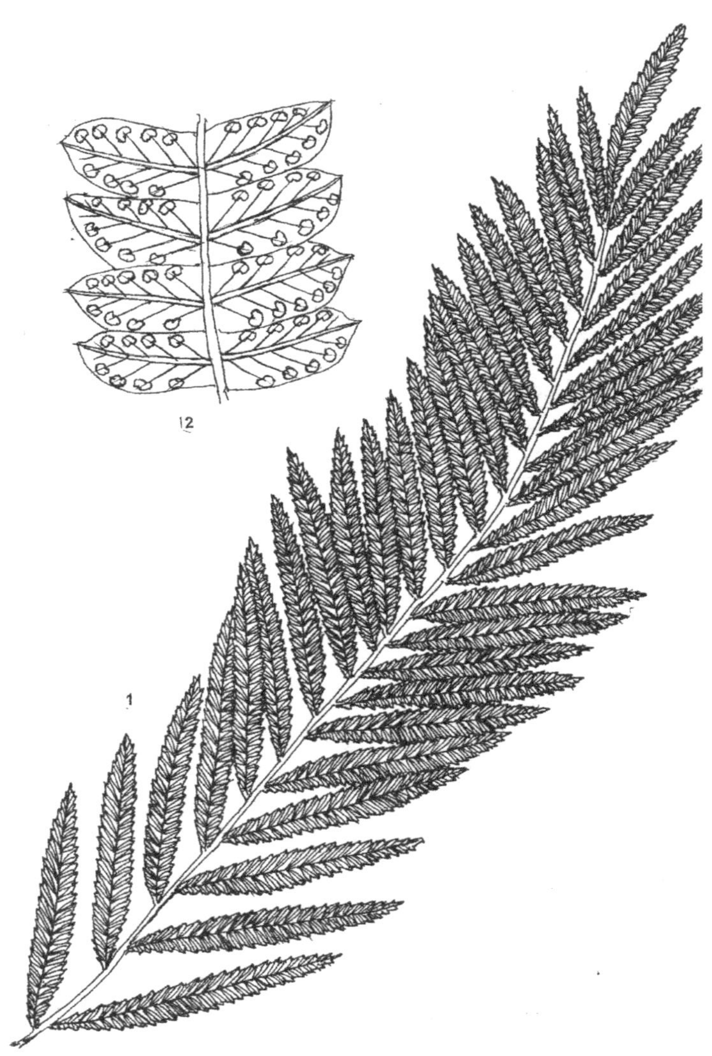

Plate – 44 *Nephrodium propingquum* Venkode plant

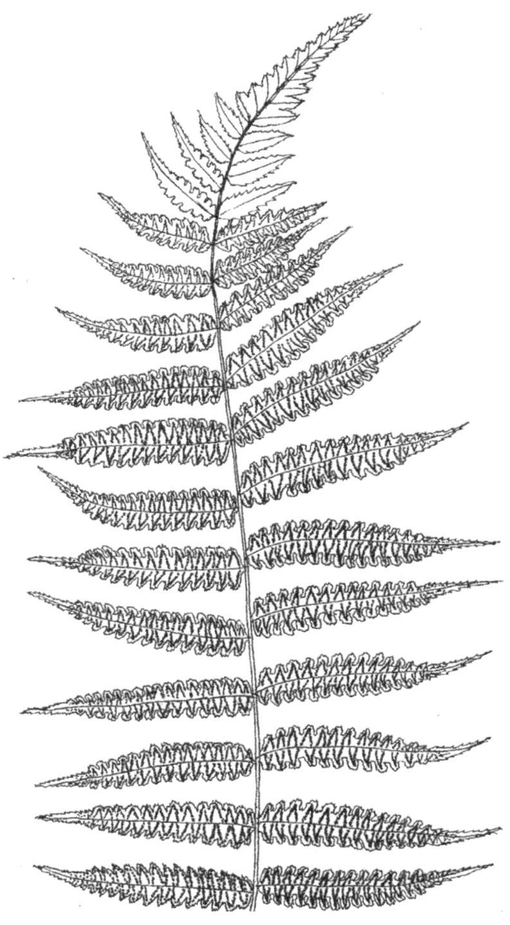

Plate – 45 *Amphineuron terminans* Venkode plant

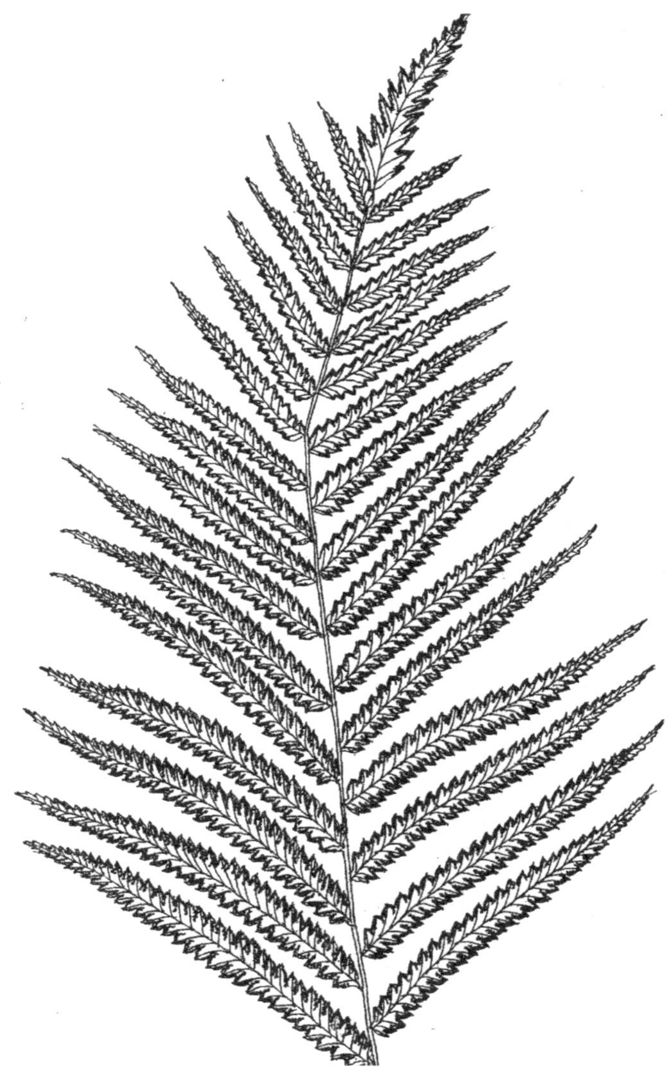

Plate – 46 *Amphineuron terminans* **Keeripparai plant**

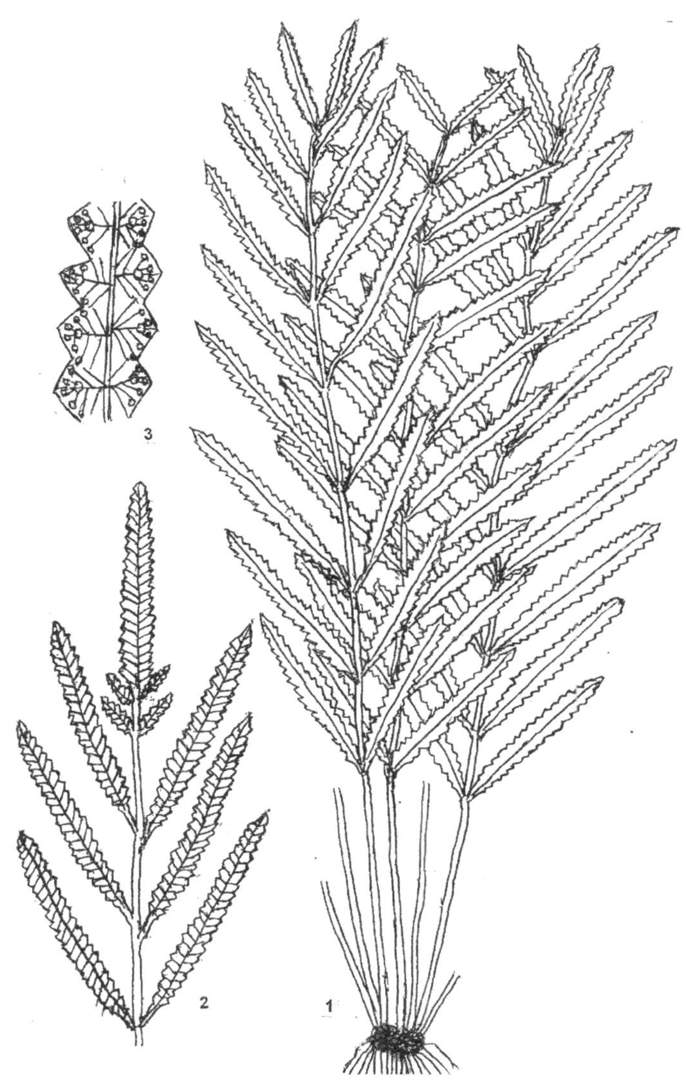

Plate – 47 *Nephrodium arbuscula* **var.** *robusta*

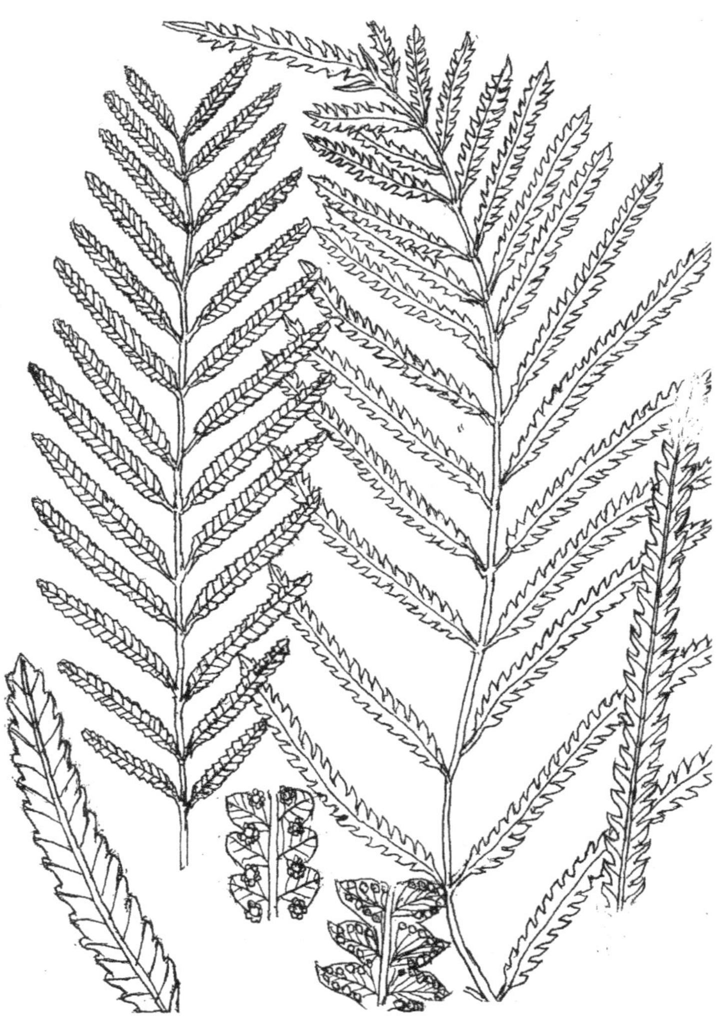

Plate – 48 **Fig.1** = *Nephrodium arbuscula* **var.** *robusta,*
Fig.2= *Amphineuron terminans* var. *flexuosa*

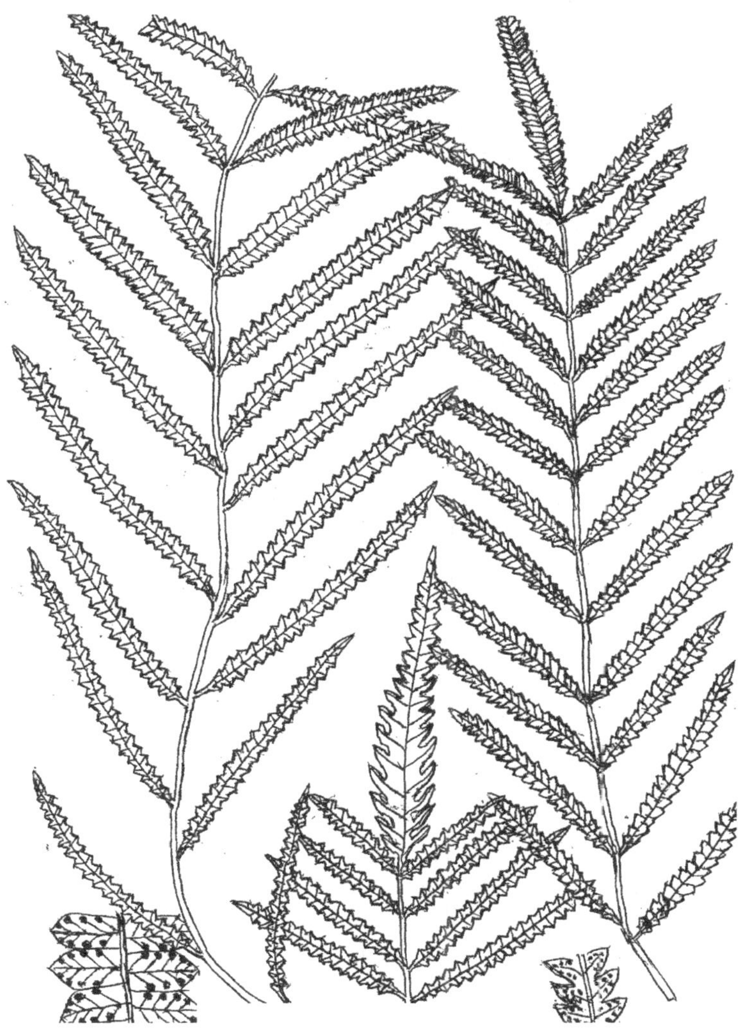

Plate – 49   Fig.1 =*Amphineuron terminans* var. *elongata*, Fig. 2 = *Amphineuron terminans* var. *nephrolioides*

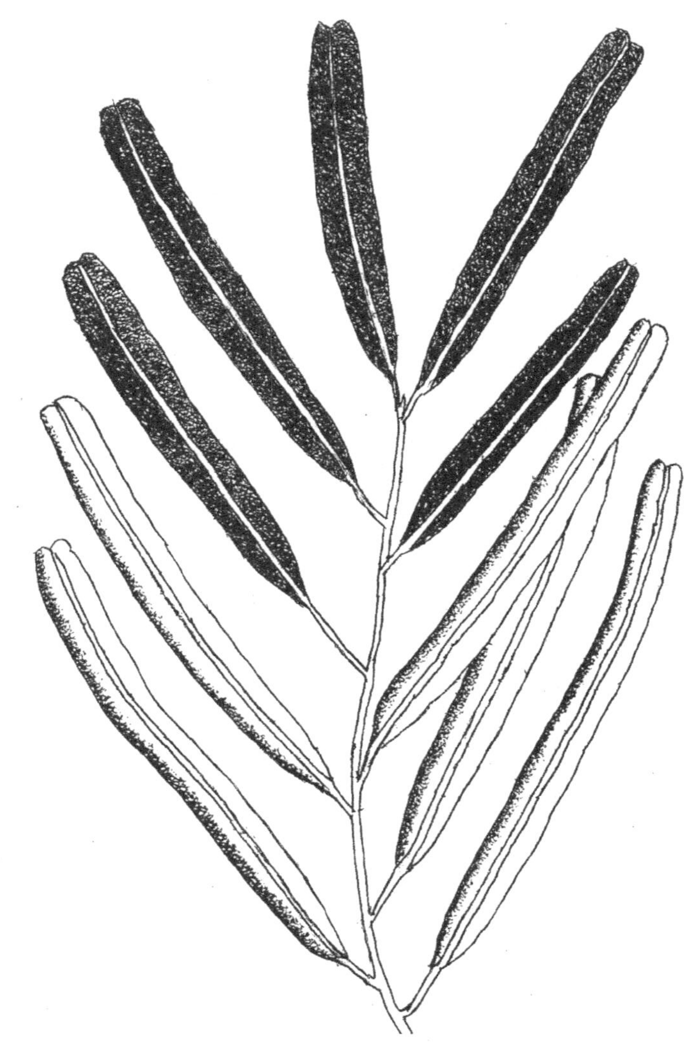

Plate – 50 *Acrostichum aureum* **from Kanyakumari**

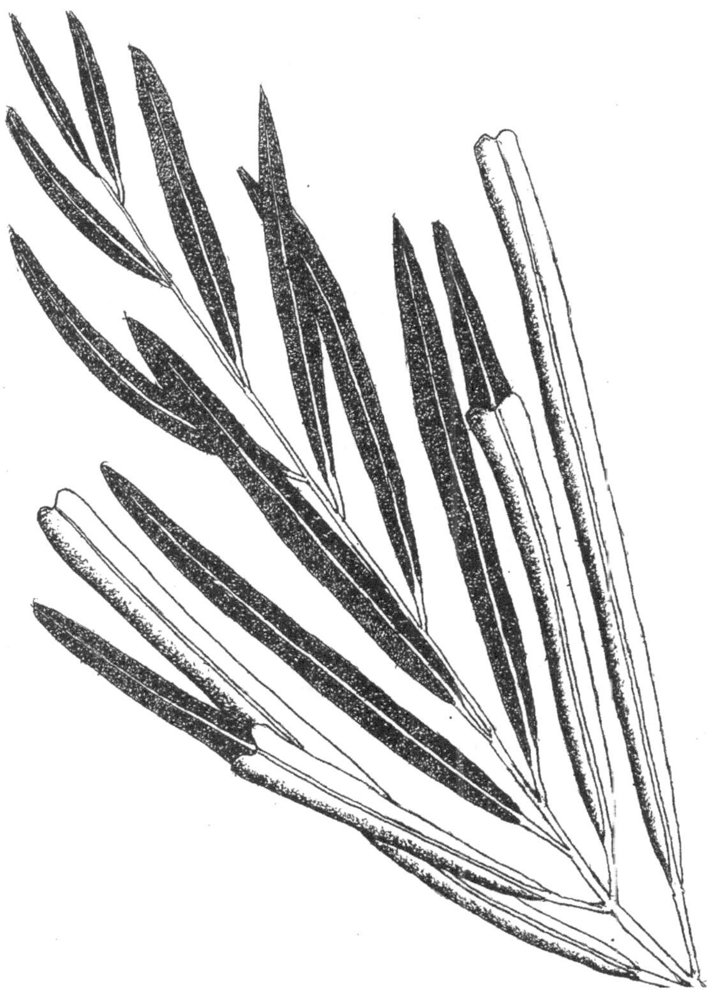

Plate – 51 *Acrostichum aureum* **var.** *acuminatum* **from Neyveli**

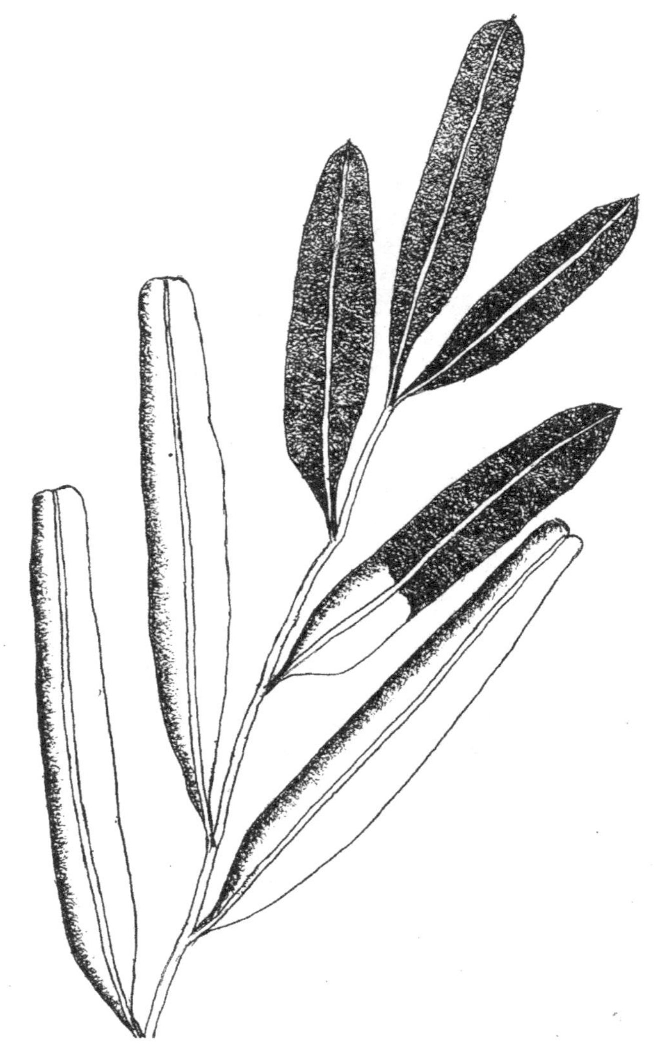

Plate – 52 *Acrostichum aureum* **Goa- variety**

Plate – 53 *Acrostichum aureum* **Kochi variety**

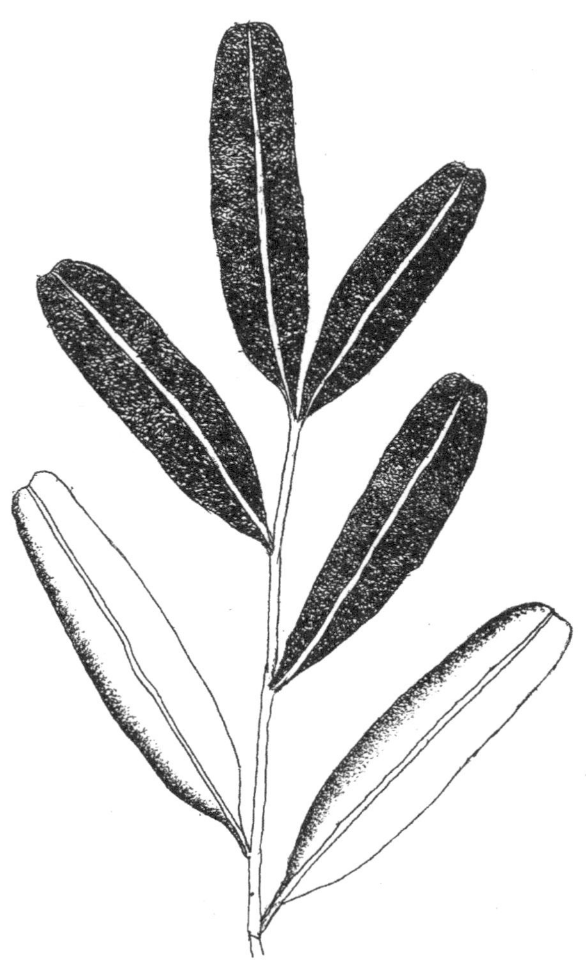

Plate – 54 *Acrostichum aureum* **from Olavanna, Kozhikode**

Plate – 55 *Ceratopteris thalictroides* **from Cuddalore.**

Plate – 56 *Ceratopteris thalictroides,* Fig.1 = Collected from Kanyakumari, Fig.2 = Collected from Palakkode

Plate – 57 *Pityrogramma calomelanos* Silver fern from Neyveli, 2 different varieties,

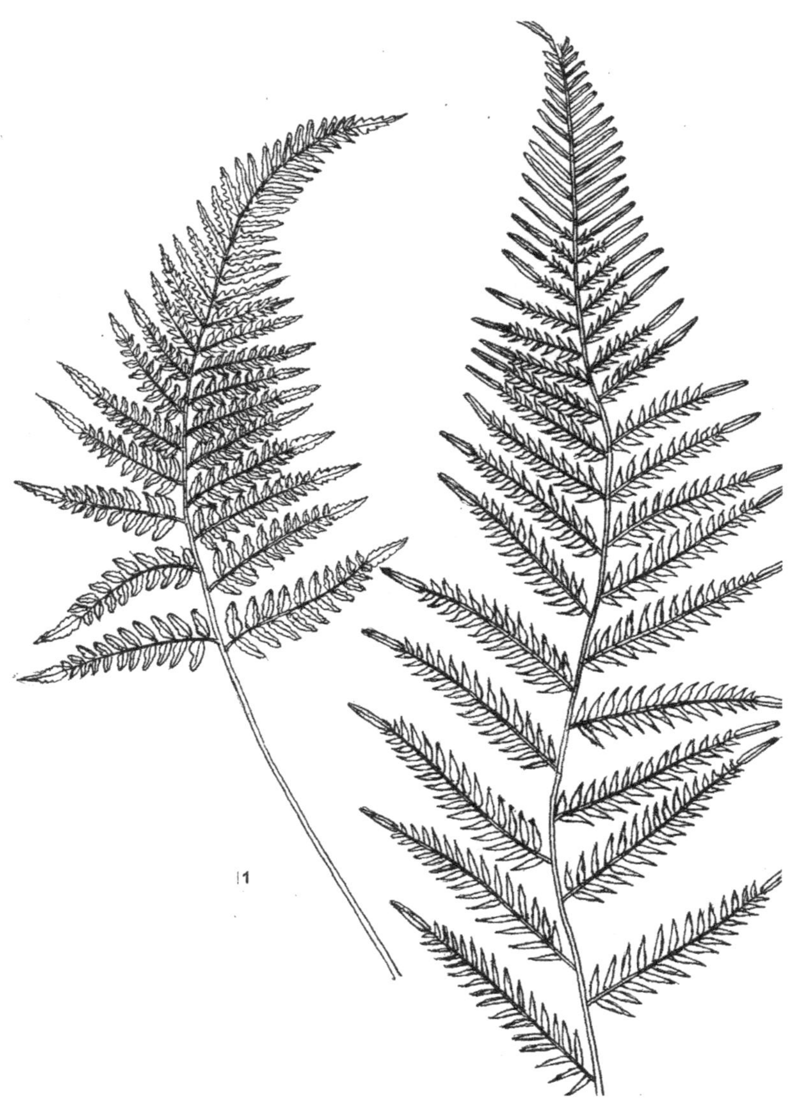

Plate – 58 *Pityrogramma calomelanos,* Fig.1= From Kodaikkanal Fig.2= From Kambam

Plate – 59 *Pityrogramma chrysophylla* = Gold fern. Fig.1 = From, Ketti, Ooty, Fig.2 = From Yercaud.

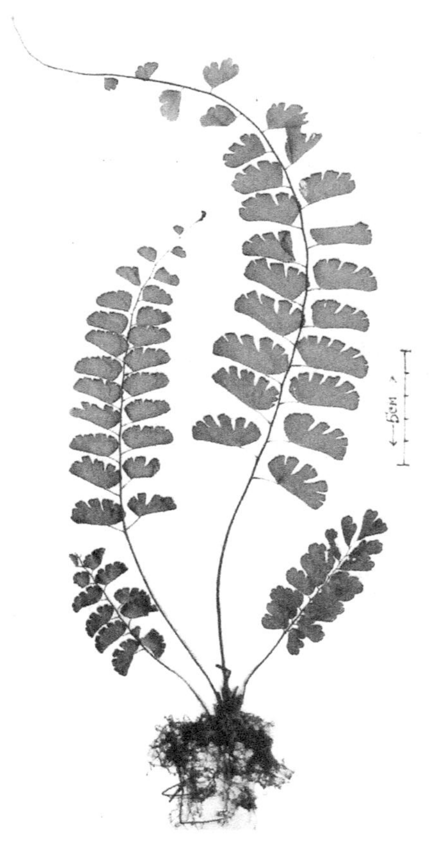

Plate – 60   *Adiantum hispidulum*

Plate – 61 *Adiantum incism*

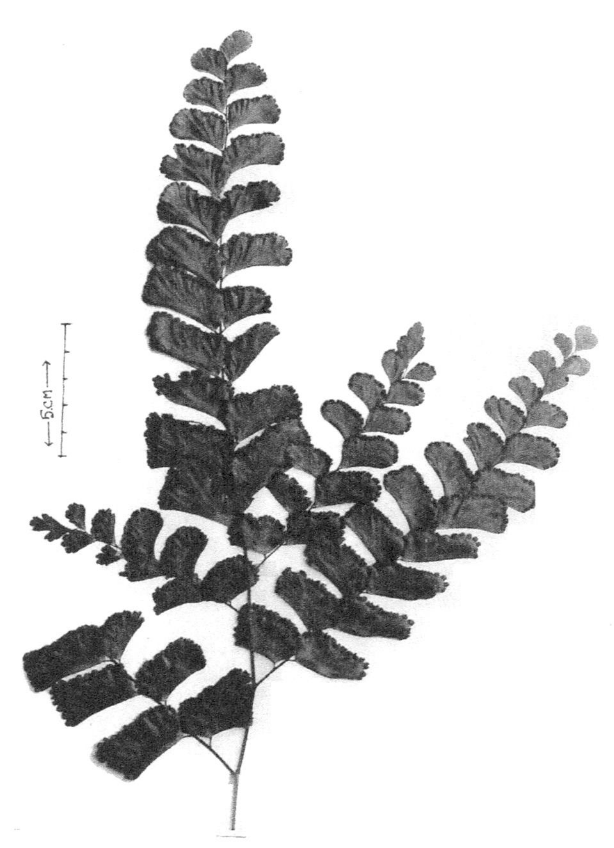

Plate – 62 *Adiantumseemannii*

Plate – 63 *Adiantum tenerum*

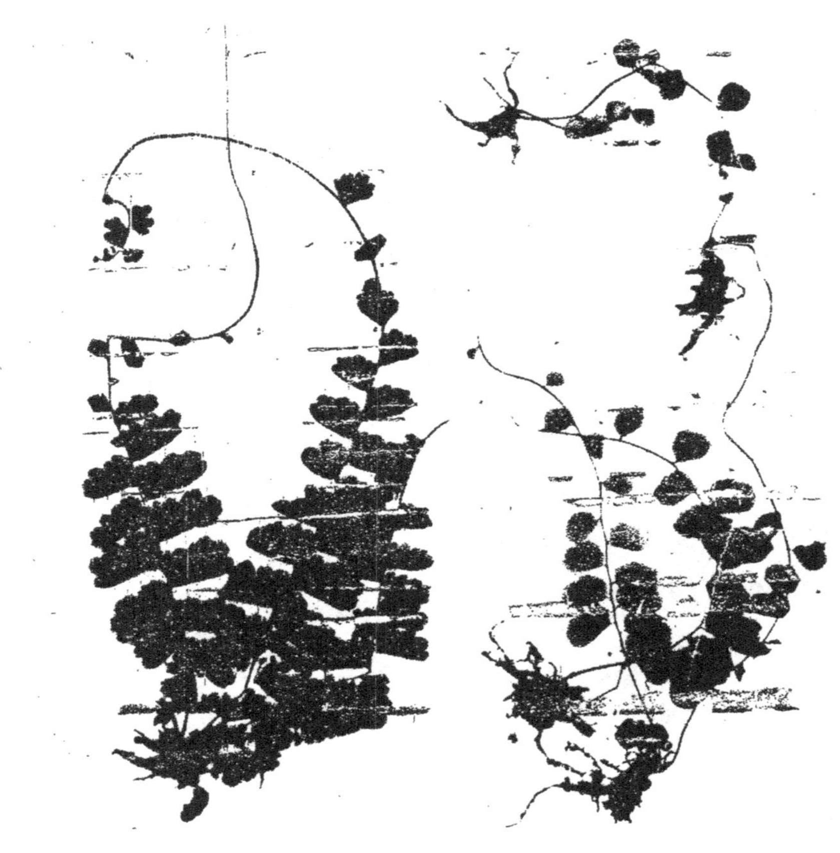

Plate – 64 A= *Lindsaea repens.* var *minor;* B = *Lindsaea repens* var. *rotundifolia*

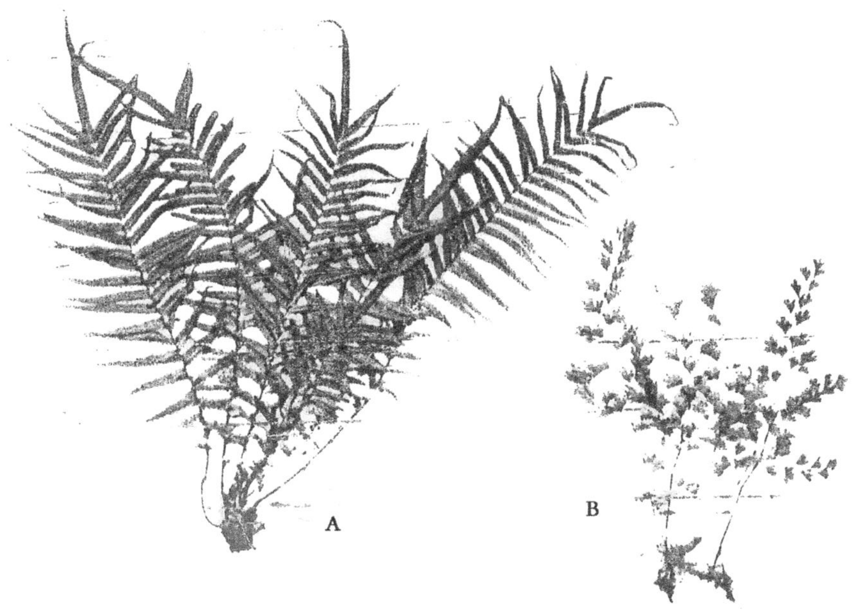

Plate – 65 A= *Pteris vittata* from Athanur, B = *Adiantum capillus – veneris*

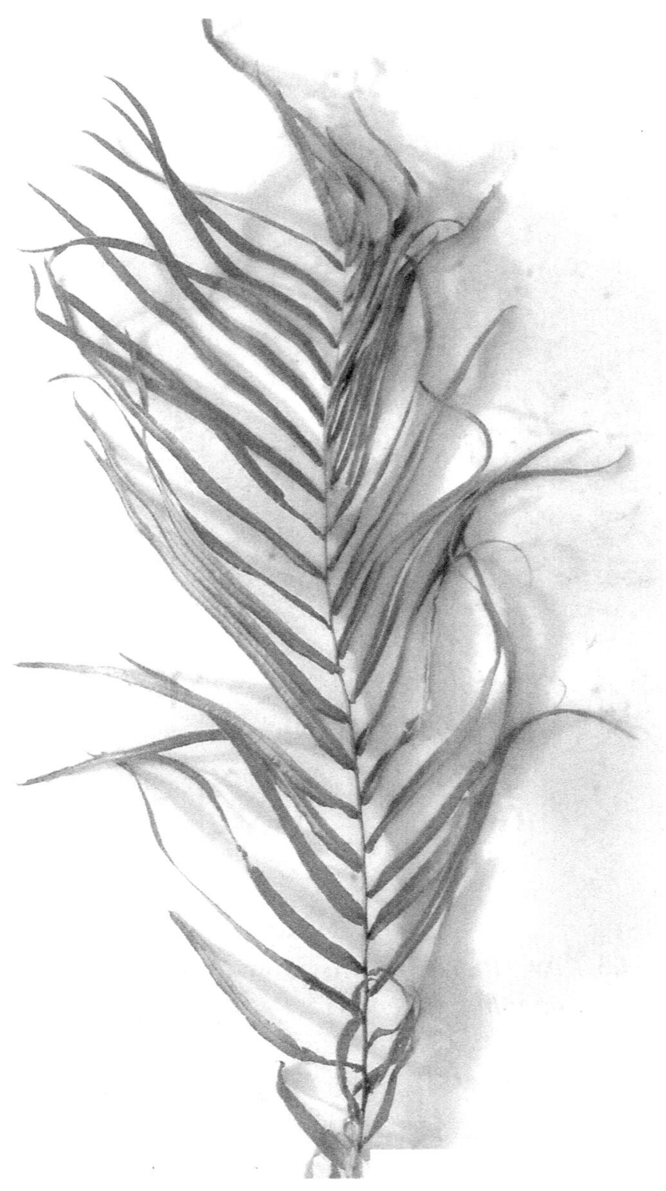

Plate – 66 *Pteris vittata* from Neyveli

Plate – 67 *Pteris vittata*from Neyveli

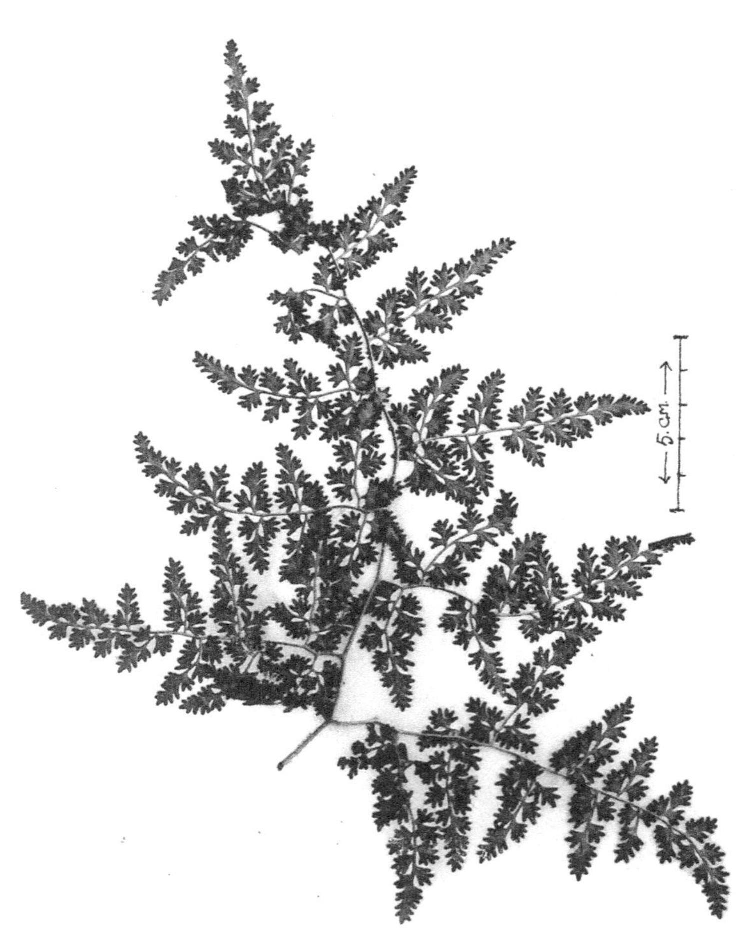

Plate – 68 *Lycodium microphyllum*

Plate – 69 *Lycodium flexuosum*

Plate – 70 *Lycodium flexuosum*

Plate – 71 *Amphineuron terminans*

Plate – 72 *Acrostichum aureum* from Neyveli

Plate – 73 *Acrostichum aureum* from Neyveli; Branches with fertile leaves at the tip.

Plate – 74 *Ceratopteris thalictroides* from Cuddalore

Plate - 75 *Pityrogramma calomelanos* = silver fern. From Neyveli, upperside

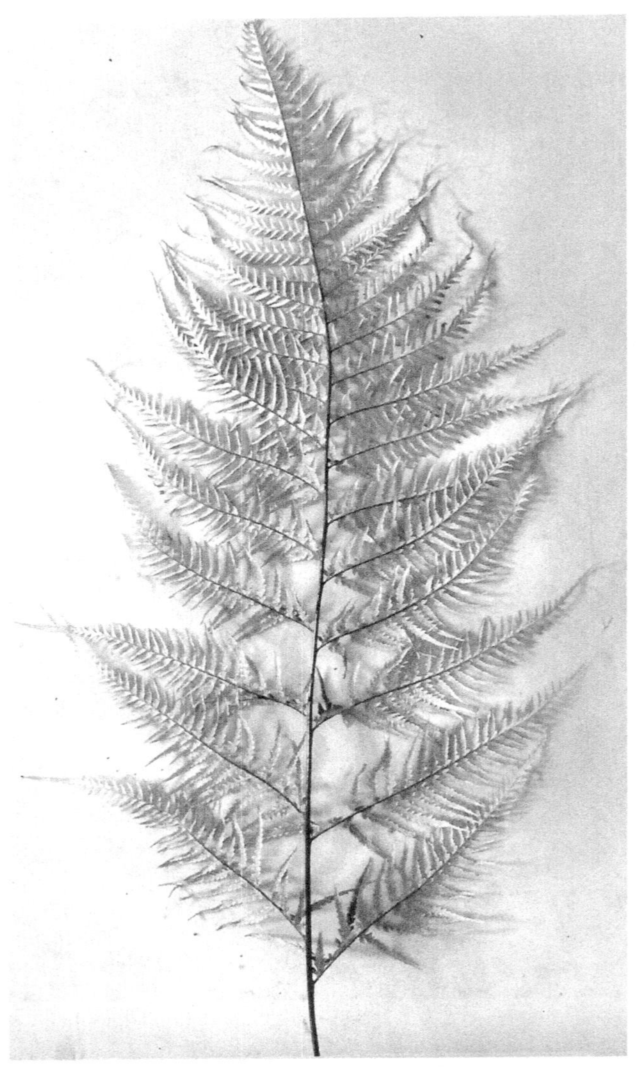

Plate – 76 *Pityrogramma calomelanos* from Neyveli. Silver fern, lower side

Appendix

Table – 1 Quantitative characters of different varieties of *Pteris vittata*

S. No	Characters	Athanur plant	Chittoor plant	Neyveli plant	Poon dian kup pam platn
1	Plant height in cm	30 to35	40 to 45	70 to 80	35 to 40
2	Number of fronds	7 to 8	6 to 7	6 to 7	4 to 5
3	Pinna length in cm	4 to 7	6 to 10	18 to 23	22 to 25
4	Pinna breadth in cm	0.8 to 1.0	1.o to 1.2	1.0 to 1.2	1.0 to 1.3
5	Number of pinnae in pairs per frond.	18 to 23	20 to 25	32 to 35	15 to 18
6	Tipleaf length in cm.	7 to 8	10 to 12	12 to 15	13 to 17
7	Tip leaf breadth incm.	0.7 to 0.8	0.9 to 1.0	1.0 to 1.1	1.0 to 1.2

Table – 2 Quantitative and qualitative characters of different varieties of *Cheilanthus mysorensis*

S.No	Characters	C. mysorensis (Beddome, 1856)	Thiruvan namalai variety	Courtallum variety	Madras variety
1	Number of fronds per plant	5 to 6	12 to 14	9 to 10	5 to 6
2	Frond's length in cm	22 to 25	10 to 13	13 to 15	14 to 16
3	Frond's breadth in cm	3.2 to 3.5	3 to 3.4	3.8 to 4.0	2.8 to 3.1
4	Number of primary laterals in pairs	23 to 25	12 to 15	15 to 17	19 to 22
5	Lobes of lamina (pinna) in pairs	9 to 10	6 to 7	5 to 6	6 to 7
6	Texture of lamina	Thick	Thin	Thin	Thick
7	Frond'sshape	Narrow below broad middle narrow tip	Narrow base broad tip	Narrow base broad tip	Narrow base broad middleNarrow tip

References

1. Beddome, R.H. 1856,"*The ferns of Southern India*" Higginbotham Publishers, Madras, II edition.

2. Beddome, R.H. 1976."*Handbook to the Ferns of British India, Ceylon and Malay Peninsula with Suplement*",Today and Tomorrow'sPrinters and Publishers, New Delhi – 5, India.

3. Nayar, B.K. and K.Geevargheese 1993."*Fern flora of Malabar*", Indus Publishing Company, New Delhi,

4. Chaudhary, B.L.,Y.S.Knichi and S.Bhargava, 2006.Diversity of Bryophyte and Pteridophyte flora of Kumbhalgarh wild life sanctuary, Rajasthan, India, *Plant Archives*, 6(2); 897-813.

5. Manickam, V.S. and Irudayaraj, 1992*Pteridophytic Flora of Western Ghats, South India*, IB Publishing Pvt. Ltd, New Delhi.

6. Subramanian, D. 2007.Diversity of flora of Bryophytes and Pteridophytes from the plains of Tamil Nadu.Seminar on Recent Advances in Medicinal plants.Research, Botany Department, Annamalai University, Abs, No.1.14, p.17.

7. Subramanian, D. 2010. "Biodiversity of South Indian species of *Adiantum*"National Seminar cum "workshopon Recent Trends on plant Science Research, Botany Department, Annamalai University, Abs. No.2.53, p.44.

8. Alfrted Byrd Graf. 1981. *Tropica*"Rochrs company publisher East Ruther ford,U.S.A. Rashid, A. 1976. "*An Introduction to Pteridophyta* (Diversity and Differentiation") Van Education books, New Delhi.

Index